Remote Sensing of Environment

Remote Sensing of Environment

Ghanashyam Sharma
Aswini N K

2026
Agri Horti Press
New Delhi-110 084

ISBN: 978-93-49690-48-6 (Hardbound)

Published By

Agri Horti Press

New Delhi

E-mail: agrihortipress@gmail.com

Digitally Printed at : Replika Press Pvt. Ltd.

Preface

Remote Sensing of Environment presents a detailed and methodical exploration of the science, techniques, and applications of remote sensing as it pertains to environmental observation and analysis. In an era marked by rapid environmental change and increasing demand for sustainable resource management, remote sensing has emerged as a vital tool for collecting timely, accurate, and spatially comprehensive data across diverse ecosystems and landscapes.

This book is designed to serve the academic and professional needs of students, researchers, and practitioners in environmental science, geography, agriculture, forestry, hydrology, urban planning, and related disciplines. It begins by introducing the fundamental principles of remote sensing, including electromagnetic radiation, spectral signatures, sensor technologies, and image acquisition. These foundational concepts are essential for understanding how remote sensing captures information about the Earth's surface and atmosphere.

Subsequent chapters delve into the different platforms and sensor systems used in environmental monitoring, ranging from optical and infrared sensors to microwave and LiDAR systems. Attention is given to both satellite and airborne platforms, highlighting the strengths and limitations of each in terms of resolution, frequency, and data availability.

The core of the book emphasizes the processing, interpretation, and analysis of remote sensing data. Key topics include radiometric and geometric corrections, image classification techniques, change detection, and the integration of remote sensing data with geographic information systems (GIS). These methods form the analytical foundation for deriving meaningful insights from remotely sensed imagery.

A substantial portion of the text is devoted to practical applications, demonstrating how remote sensing contributes to the monitoring and management of forests, water bodies, soil resources, urban environments, agricultural productivity, and climate-related phenomena. Real-world case studies are provided to illustrate the role of remote sensing in environmental assessment, disaster response, land use planning, and ecosystem health evaluation.

Emerging trends and technologies in remote sensing—such as high-resolution satellite constellations, UAV-based data collection, hyperspectral imaging, and machine learning applications—are also explored. These advancements are expanding the capabilities of environmental monitoring and offering new avenues for research and operational use.

With a blend of theoretical insight and practical guidance, Remote Sensing of Environment aims to foster a deeper understanding of how remotely sensed data can be used effectively to address critical environmental challenges. The inclusion of illustrative examples, data analysis exercises, and current research highlights ensures that readers are well-equipped to engage with both foundational knowledge and cutting-edge developments in the field.

Ghanashyam Sharma

Aswini N K

Contents

Chapter-1

Satellite and Ground Measurements for Studying the Urban Heat Island Effect in Cyprus

1. Introduction

An urban heat island (UHI) is a phenomenon whereby an urban area experiences elevated air temperatures due to anthropogenic modification of the environment and is usually more evident at night. During heat waves the local effect of an UHI is superimposed on the re- gional temperature field and as a result heat stress is enhanced. Both the intensity and the spatial structure of the observed thermal contrast of the UHI depend on various parameters, such as the structure of the urban tissue, the population density and its associated heat re- lease, the land use patterns, the vegetation cover, the surface topography and relief etc. In general terms, the UHI is becoming more intense as city sizes increase. Traditional measure- ments of the near-surface UHI are based on measurements of the air temperature using ur- ban and rural weather stations or air temperature transects. Thermal satellite sensors, which primarily measure the radiance at the top of the atmosphere in the thermal infrared, retrieve the so called land surface temperature (LST) which is the temperature measured at the Earth's surface and is regarded as its skin temperature. Given that LST is different from the surface air temperature, a distinction is made in remote sensing studies between surface ur- ban heat island (SUHI) and atmospheric heat island (e.g., Nichol, 1996).

Several studies published in the literature have focused on the use of remotely sensed data for studying the urban heat island effect (Dousset & Gourmelon, 2003; Kato & Yamaguchi, 2005; Lo & Quattrochi, 2003; Streutker, 2002; Tran et al., 2006; Xiao et al., 2007; Yuanbo et al., 2007). Other relevant studies are focusing on the validation of satellite LST retrievals with ground measurements or on the inter-comparison of LST products from different sensors (Mostovoy et al., 2005; Nichol et al., 2009; Retalis et al., 2010). The availability of a multitude of data archives (e.g., from MODIS, ASTER and Landsat TM/ETM+ sensors) with long time- series has recently raised the scientific interest in the relevant field. As a result, several stud- ies have been published on the study of the UHI effect for various cities of the world (Hung et al. 2006; Imhoff et al., 2010; Peng et al., 2012).

This Chapter discusses the urban heat island effect in Cyprus based on both multi-temporal satellite and meteorological data. The necessary information of the study area is provided in Section 2. The description and selection of the heat waves and the analysis of the synoptic conditions favouring the development of heat waves are discussed in Section 3. The devel- opment of a Neural Network for the correlation of satellite derived land surface temperature (LST) with ground based air surface temperature is examined in Section 4. The analysis of satellite derived LST for studying the temporal evolution of LST and the deviation of LST (anomaly) from the mean values during a heat wave event are presented in Section 5, while Section 6 refers to the calculation of the mean monthly magnitude of urban heat island (UHI) for the period 2002-2008 and for selected heat wave events.

2. Study area

The island of Cyprus is located in the eastern part of the Mediterranean Basin. The island is situated between latitudes circles 34° and 36° N, and meridians 32° and 35°E. Cyprus has a typical eastern Mediterranean climate: the combined temperature–rainfall regime is charac- terized by cool-to-mild wet winters and warm-to-hot dry summers (Michaelides et al., 2009). The climatological annual precipitation of Cyprus is around 500mm. The highest precipita- tion is recorded in the mountainous areas with 1100mm, while in the coastal areas precipita- tion is limited to 300-350mm.

From a morphological point of view, the island can be divided into five main morphological regions: (a) The mountainous complex of Troodos located at the center of Cyprus; (b) the mountain range of the Pentadaktylos at the northern part; (c) the central plain of Mesaoria located between of these two mountainous ranges; (d) the hilly areas around the mountain- ous complex of Troodos; and (e) the coastal plains (see Fig. 1). The coastline of Cyprus is characterized by numerous capes and bays. The narrow coastal plains in the north are cov- ered with olive trees and carob trees, while a short distance from the coast, the northern mountain range (Pentadaktylos) is found, which is a limestone formation and peaks to a height of 1024 meters. At the south and the east of the island there are two salt – lakes.

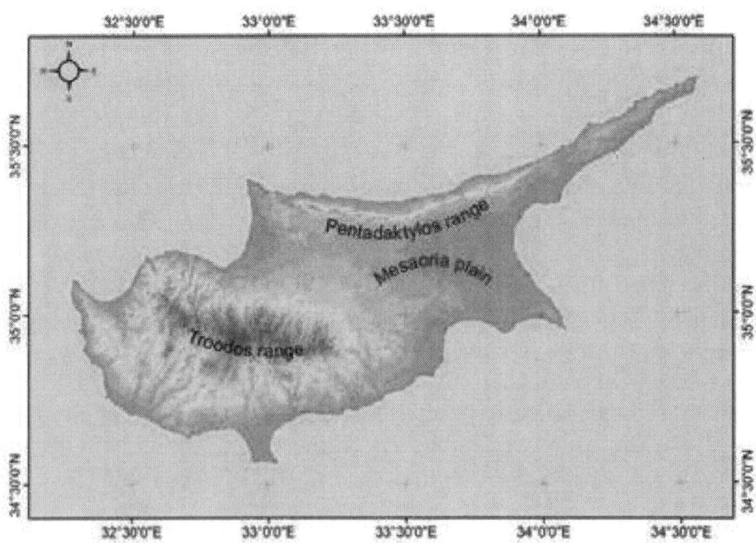

Figure 1. Main morphological regions of Cyprus

The Troodos mountain range with a peak at 1951 m covers most of the south-western part and the center of the island. This area is covered almost by forests, mainly pine and other forest trees such as cypresses, oaks and cedars. It is estimated that forests cover about 19% of the total area of the island.

Cyprus is divided into six districts: Kyrenia, Famagusta, Larnaca, Limassol, Paphos and Nicosia (Fig. 2). During the last decade (2000-2010), there has been recorded a dramatic urban expansion (see Fig. 3). As it was found from previous studies (Hadjimitsis et al., 2011), there has been an increase of urban areas of more than 100% compared to late 1980's and a decrease of 20% of rural areas. These results were derived from an analysis of multi-tempo- ral satellite image classification.

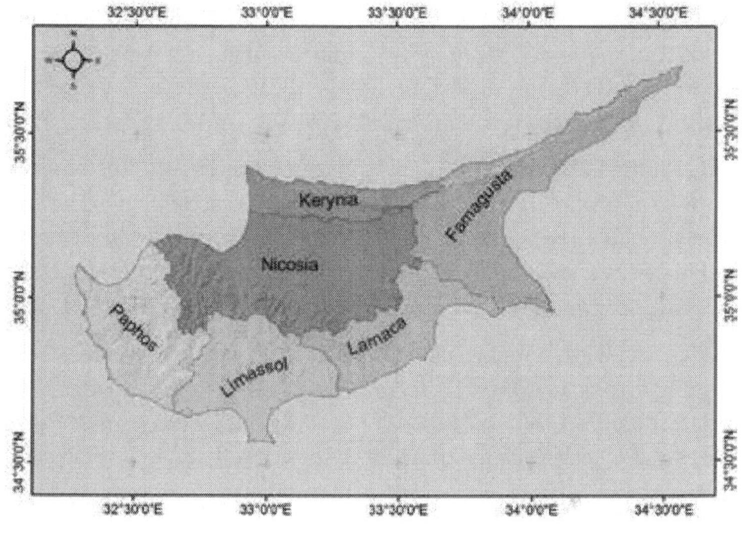

Figure 2. Districts of Cyprus

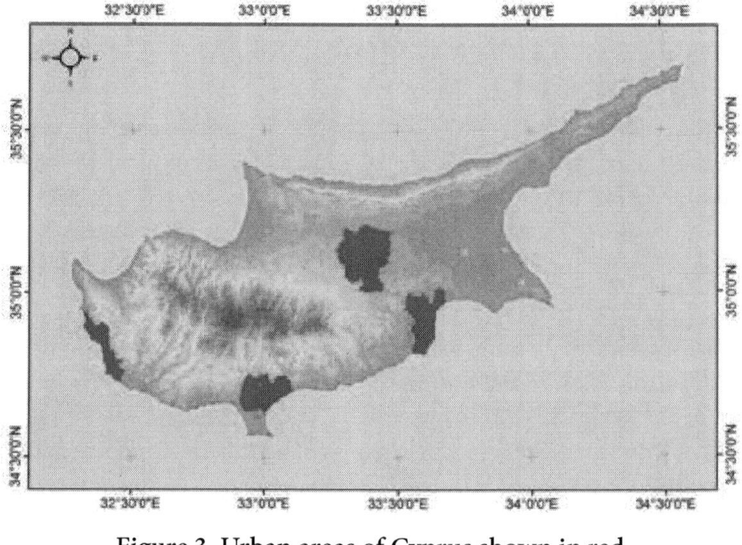

Figure 3. Urban areas of Cyprus shown in red

3. Heat wave events and synoptic patterns

A strong relationship exists between large scale circulation patterns and regional surface variables such as surface pressure, dynamical rainfall, wind and temperature (Tymvios et al., 2007, 2008, 2010a; Xoplaki et al., 2003). As a consequence, synoptic upper air charts at certain levels comprise a valuable tool for the operational weather forecaster to qualitatively predict occurrences of weather phenomena observed on the ground (e.g., heavy rainfall; see Tymvios et al., 2010a). The height pattern at 500 hPa is often used for this purpose. In order to take advantage of these semi-empirical methods and to simplify the statistical processing, stochastic downscaling methods are often applied to the actual weather patterns in order to generate clusters of synoptic cases with similar characteristics. Weather type classifications are simple, discrete characterizations of the atmospheric conditions and they are commonly used in atmospheric sciences. For a review of various classifications, including their applications, refer to Key & Crane (1986), El-Kadi & Smithoson (1992), Hewitson & Crane (1996) and Cannon & Whitfield (2002).

Heat waves have a distinct impact on society through increased mortality, change in the energy consumption profile and the diversification of social behavior. The severity of the heat events may include the local climatological characteristics, the community design and the individual tolerance to heat. Both the frequency of appearance and the intensity of heat waves are increasing in the Mediterranean area (Founda & Giannakopoulos, 2009).

The eastern Mediterranean climate is characterized by the succession of a single rainy season (November to mid-March) and a single longer dry season (mid-March to October). This generalization is modified by the influence of maritime factors, yielding cooler summers and warmer winters in most of the coastal and low-lying areas. Visibility is generally very good. However, during spring and early summer, the atmosphere is quite hazy, with dust transferred by the prevailing south-easterly to southwesterly winds from the Saharan and Arabian deserts, usually associated with the development of desert depressions (Michaelides et al., 1999). The influence of synoptic types on the urban heat island has been investigated by Mihalakakou et al. (2002) who have also adopted a neural network approach.

The definition for a heat wave recommended by the World Meteorological Organization is "when the daily maximum temperature of more than five consecutive days exceeds the maximum temperature normal by 5°C". Nevertheless, in most countries, the definition of extreme heat events is based on the potential for hot weather conditions to result in an unacceptable level of adverse health effects, including increased mortality. Also, a threshold in maximum temperature is in practical use in many countries.

These periods of abnormally and uncomfortably hot and (usually) humid weather are very common in the eastern Mediterranean during summer and early autumn. Expert examination of the synoptic patterns on upper air charts may reveal the potential for a heat wave event. In this respect, the research presented here attempts to identify height patterns favorable for heat events by using a neural network classification method, namely, Kohonen's Self Organizing Maps (see Kohonen, 1990).

3.1 Data

As an indication of a possible heat event, the maximum temperature of Nicosia station in Cyprus was chosen. This station is located within the urban area of the city of Nicosia (35°17'N, 33°35'E, 170m, see Fig. 4) and equipped with traditional instrumentation was operational from 1957 until 2001, when it was upgraded to an automatic station. The database used in this study comprises the maximum and minimum temperature records from this station. The maximum monthly temperature measurements are presented in Fig. 5. Also, for the classification of synoptic patterns, the ERA40 reanalysis for the period of 1958 to 2000 (covering roughly the ERA40 time window) were utilized.

The temperatures database was checked for consistency and homogeneity against measurements from nearby stations while the maximum temperatures were also checked for normal distribution fitting.

3.2 Methodology

The maximum daily temperature at Nicosia station was checked against the climatological monthly average value of the period 1961-1990. If the difference was 5°C or more, then the period was characterized as "possible heat event". If the subsequent days were also positive against this temperature test for more than three days, then the period was considered as heat event. The heat events were checked against the weather classification patterns in order to identify a connection among particular patterns and heat events. The same procedure was adopted for a difference of 3°C, since events with a 5°C difference are rare even during summer. Special care was taken when checking the last and the first day of the month whereby daily maximum temperature values were subtracted from the average climatological value of the two subsequent months.

Figure 4. Location of ground stations used

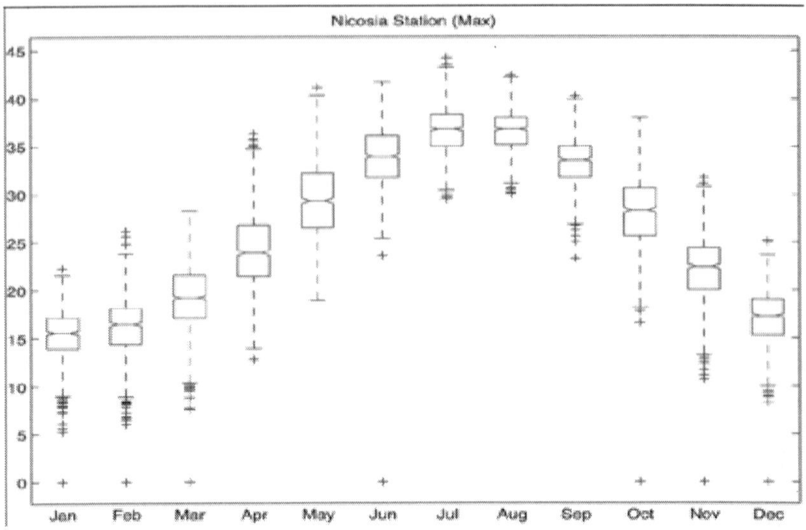

Figure 5. Box and Whiskers plot of the maximum temperatures in Nicosia (1958 - 2000)

Details of the Self Organizing Maps methodology used for the classification have been presented in Michaelides et al. (2010). The 36-Cluster classification adopted also in the present study has been recently demonstrated by Tymvios et al. (2010b).

3.3. Results

The distribution of the heat events in consecutive days for 3°C and 5°C difference is illustrated in Fig. 6. It is clearly evidenced that more than 75% of the events last for 3 to 5 days. Most of the identified heat events occur in the transition periods (i.e., Spring and Autumn). This finding is also supported by the findings in Fig. 5, where the larger variation (the area be- tween 25th and 75th percentile) of the average of the maximum temperatures is given for the same periods. With the exception of the periods 12 to 21 July 1978 (10 days) and 2 to 14 July 2000 (13 days), all incidents lasting more than 10 days for this station occurred in October, November, March, April and May.

Clusters 5 and 34 share most of the heat event occurrences. They are both transition period clusters with similar characteristics, exhibiting a distinctive upper level ridge over the eastern Mediterranean and a deep low to the west of this ridge; Cluster 5 belongs to the cold period and Cluster 34 to the warm period. An example of a Cluster 5 member is illustrated in Fig. 7.

When these clusters appear during early Spring and late Autumn, the heat events last from 8 to 15 days, while when they appear just before or after Summer (May and September) they last around 5 days. Summertime appearances of heat events are equally shared between Clusters 12, 19, 24 and 36, all characterized by warm and dry conditions (Michaelides et al., 2010).

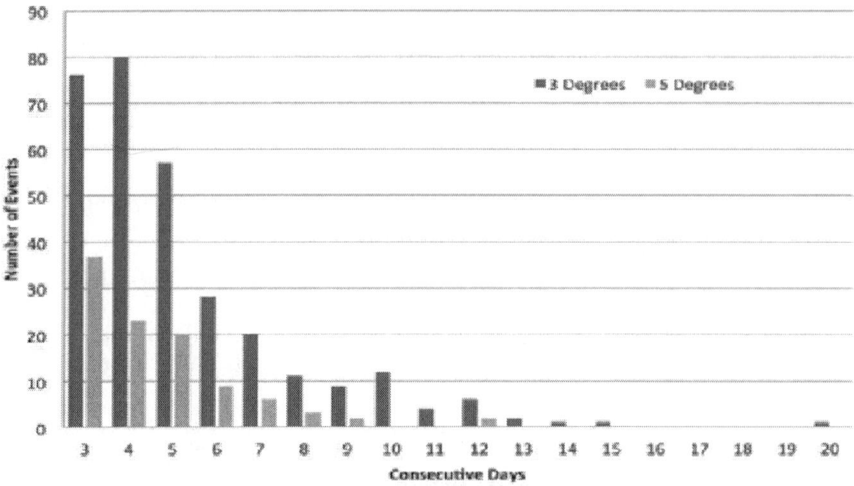

Figure 6. Distribution of the heat events in consecutive days for 3°C and 5°C difference

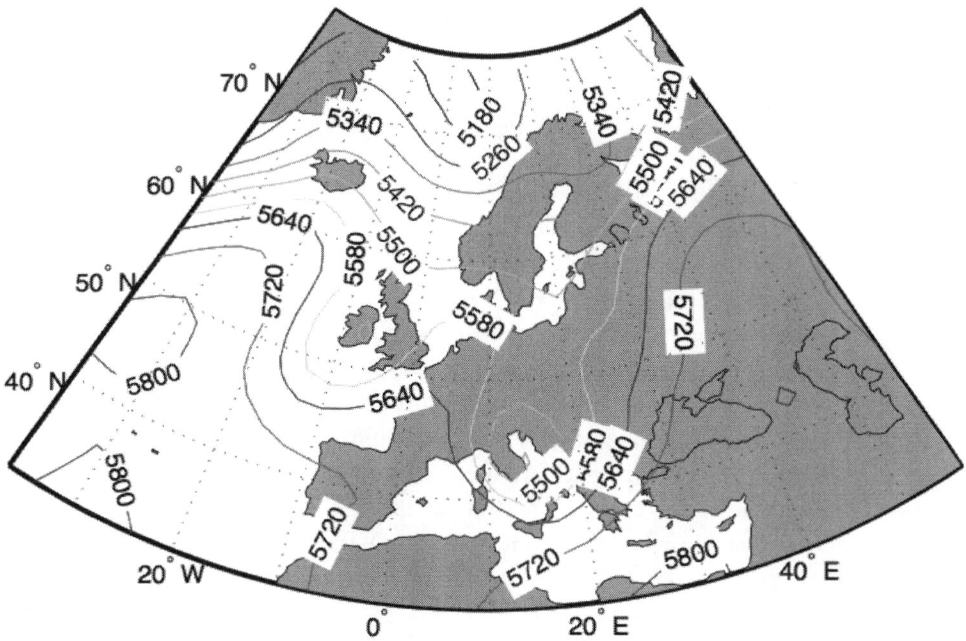

Figure 7. The height pattern at 500hPa (Cluster 5) from 24 November 1962

Although it appears that some Clusters are associated with heat events over Cyprus, the connection between heat events and atmospheric circulation at 500hPa did not give definite results that any of these patterns dominate heat event occurrences (as it was possible to demonstrate in previous studies on rainfall and extreme rainfall events). There might be two reasons for this inadequacy. The first is that the window that was chosen for the classification does not include the synoptic patterns that influence the area sufficiently;

the second reason is that, although upper air patterns at 500hPa contribute significantly to the evolution of certain surface features (such as dynamical or extreme rainfall), such an association is not so clear for the temperature field. In the search for associations of the temperature fields with synoptic patterns in the Mediterranean, it is important to consider also the lower parts of the atmosphere.

Future research concerning the connection of the weather classification patterns will be focused into a new, much larger window that will include Northern Africa and the Middle East and a combination of classification of patterns at lower levels of the atmosphere (e.g., 850hPa, 700hPa).

4. Satellite estimates of temperature versus ground measurements

In this Section, a methodology is presented in which the temperature estimates from the MODIS sensor onboard the Terra Satellite is contrasted with ground measurements. The methodology consists of a neural network approach in which measurements on the ground are used as input to the neural network, whereas, the temperature estimate from the satellite is considered as the network's output.

The neural network methodology adopted has successfully been implemented before in tackling several climatological problems in Cyprus: the prediction of maximum daily total solar irradiance (Kalogirou et al., 2002), the prediction of the daily average solar radiation (Tymvios et al., 2002, 2005a), the modeling of photosynthetic radiation (Tymvios et al., 2005b) and others.

4.1 Data

For the needs of this research, data from MODIS onboard the Terra satellite have been used. More specifically, the level-3 product MOD11A1 (version 5) for the period 2000-2009 was exploited, at a resolution of about 1km by 1km (0.927km). Using the available Land Surface Temperature (LST) fields derived from MODIS, a time series was established corresponding to the position of ground stations. Wan & Dozier (1996) have developed the Generalized Split Window (GSW) algorithm for the retrieval of LST, using the thermal (infrared) channels of MODIS and under different atmospheric conditions (see also, Wan, 1999, 2008). This algorithm retrieves LST on the basis of emissivities in bands 31 and 32 of MODIS. The accuracy in estimating LST was found to be better than 1K, whereas in most cases it was better than 0.5K (Hulley & Hook, 2009; Coll et al., 2009).

The data base for the surface measurements used in this research consist of the hourly recorded temperature at each of the automatic meteorological stations of the network operated by the Cyprus Meteorological Service (see Fig. 4), in the period 2000-2007. Based on these data, the maximum temperature recorded in the time period 1100 – 1300 UTC (local standard time=UTC+2 hours) was considered as the day's maximum and was subsequently used in the study.

The training of the neural network implemented requires that there are no missing data in the time series used, because the data are used in groups and are therefore interdependent. Therefore, the estimated LST (by the neural network implemented) is based on the data of a whole day and missing values result in the rejection of that day. Following quality control based on the above constraint, the number of automatic stations that were subsequently used was reduced to twelve, as shown in Table 1.

4.2 Methodology

Artificial Neural networks (ANN) are small autonomous computational units (algorithms) with inter-connections which, to a large extent, resemble the functioning of natural computational units, namely, the neurons of the human brain. ANN can be trained and learn through repeated examples so that they can reach conclusions and results without human intervention. Since their invention, ANN covered a wide spectrum of research and disciplines and their application has been phenomenal. A few of the numerous examples of ANN applications are mentioned here: medical systems' automation for the recognition of malignant tumors, control of military equipment and aircraft, estimation of environmental variables, quality verification in production factories, forecasting of financial indices, weather diagnosis and forecasting etc.

For the implementation of the ANN methodology in the present research, the Multi-Layer Perceptron (MLP) was adopted (see Haykin, 1998). The input to this network is the surface temperature recorded at the ground stations and the output is the temperature estimated by the satellite (LST). Fig. 8 displays the MLP implemented.

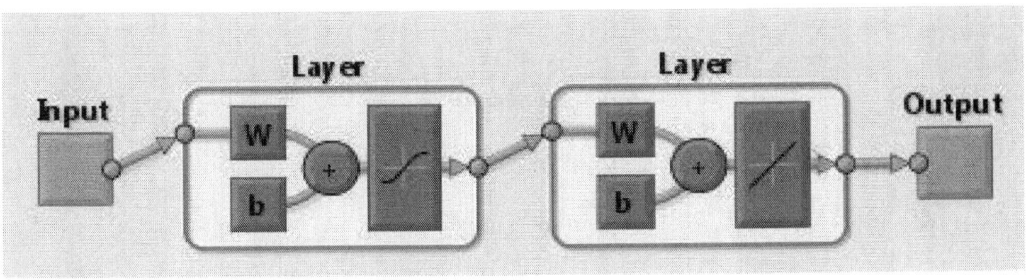

Figure 8. The Multi-Layer Perceptron (MLP) network implemented for the prediction of LST

The data from the twelve ground stations and the respective MODIS estimations of LST were used as follows: 60% were used for Training of the network, 20% for Validation and the remaining 20% as an independent set for Testing.

4.3. Results

Table 1 displays the errors in the estimation of LST with the neural network, by using the independent set of data. In this table, the maximum, minimum and average errors along with the standard deviation are shown for each of the ground stations. Overall, the performance of the neural network is considered as very satisfactory. However, there are cases where the error is unacceptable and this requires further investigation.

The relation between input data (ground temperature) and satellite estimated temperature LST (target) is shown in Fig. 9. The results are shown for all the data but also separately for the Training, Validation and Testing data sets. For the Training set, the correlation coefficient is R=0.96991, for the Validation set R=0.89692 and for the Testing set R=0.9145, whereas, for all the data R=0.94747. Based on these findings, the performance of the network in predicting LST is considered as satisfactory.

Table 1. Errors of LST estimation for the independent set of data for each ground station

Ground station	Latitude (N)	Longitude (E)	Altitude Above sea level (m)	Maximum (°C)	Minimum (°C)	Average (°C)	Standard deviation (°C)
Astromeritis	35°03′	32°26′	175	6.81	-7.31	0.22	2.35
Athalassa	35°04′	33°58′	162	7.14	-7.58	0.63	2.77
Athienou	35°03′	33°32′	185	6.45	-5.74	0.23	2.27
Dasaki	35°03′	33°47′	50	5.74	-4.52	0.60	2.17
Kannaviou	34°55′	32°35′	419	7.03	-6.23	0.48	2.41
Kathikas	34°55′	32°26′	650	8.25	5.30	0.63	2.53
Kato Pyrgos	35°11′	32°41′	5	7.19	-10.12	0.62	2.99
Malia	34°49′	32°47′	645	7.59	-7.75	-0.11	2.56
Mennogeia	34°51′	33°26′	140	7.36	-7.78	0.28	3.07
Paphos	34°47′	32°26′	82	8.34	-7.45	0.37	2.83
Paralimni	35°04′	33°58′	65	7.51	-8.42	0.68	2.61
Polis	35°03′	32°26′	20	6.87	-4.83	0.40	2.40

In this research, an attempt has been made to relate ground measurements of temperature with the temperature as it is estimated from MODIS and develop a neural network method- ology that can be used in the estimation of ground temperatures by using the satellite im- agery. Although the methodology performs sufficiently, overall, it seems that further refinement is needed in order to improve the approach. The adoption of a single network for all the time series of data seems to limit the application of the methodology. For example, the present single neural network developed for each station does not take into account the large seasonal variations in the parameter concerned. It could be more effective if several neural networks are developed based on seasonally grouped data.

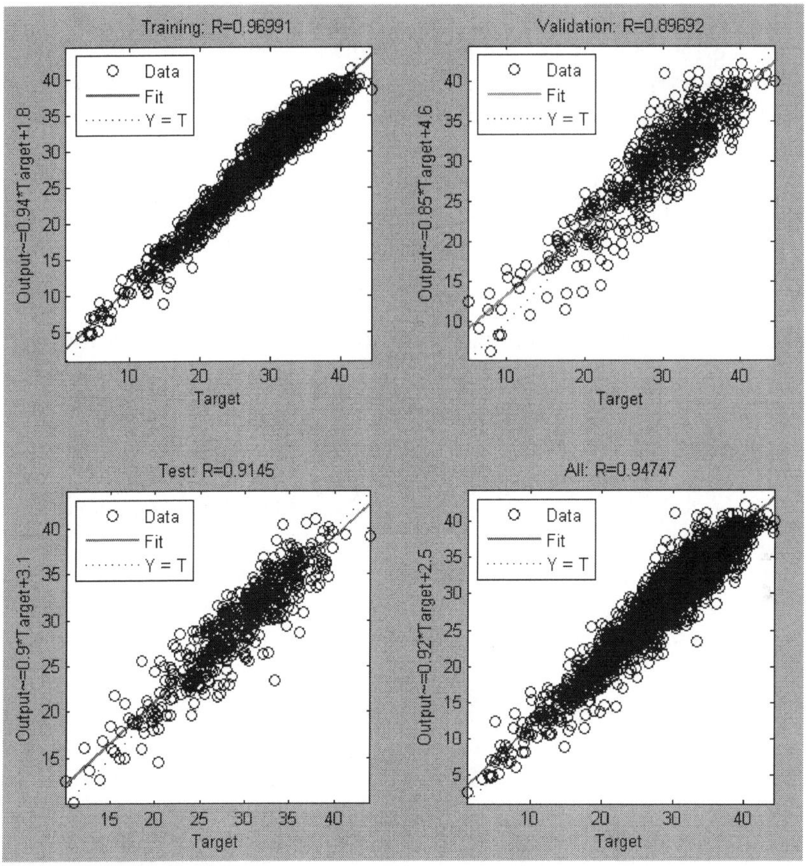

Figure 9. Neural network performance for the Training, Validation, Test and All data sets

5. Land surface temperature analysis

The MODIS sensor, onboard Terra and Aqua polar satellites, provides one day and one night image under clear sky conditions. MODIS is particularly suitable for the land surface temperature (LST) product due to its global coverage, radiometric resolution and dynamic ranges for a variety of land cover types and high calibration accuracy in multiple thermal bands.

MODIS LST product is based on the generalized split-window (GSW) algorithm (Wan & Dozier, 1996) using as input the MODIS thermal bands 31 and 32. The parameters in the MODIS GSW depend on the satellite zenith view angles, column water vapor and also on the low atmosphere boundary temperature. The band emissivities rely on the classification- based method (Snyder et al., 1998) according to land cover types in the pixel (Monteiro et al., 2007). Temperatures are extracted in Kelvin; accuracy of 1 Kelvin is yielded for materials with known emissivities (Wan, 1999), while a number of studies have also tested the accuracy of the MODIS LST product with favorable results (Wan, 2002; Wan et al., 2004; Coll et al., 2005; Wan, 2008).

The MODIS Aqua product MYD11A1 (V5) and MODIS Terra product MOD11A1 (V5) – Land Surface Temperature and Emissivity Daily L3 Global 1 km Grid SIN were used. Terra and Aqua overpass times for the study area are considered at approximately 1030 and 1330 UTC for day passes, and at approximately 2230 and 0130 UTC for night passes, respectively.

The use of MODIS LST data for examining the temporal evolution and the retrieved temperature anomaly maps for a heat wave event occurred on 24 June 2007 is presented. Moreover, MODIS LST data are used for calculating the urban heat island (UHI) at four urban areas of Cyprus during the extreme heat wave of August 2010.

5.1. MODIS LST temporal evolution and temperature anomaly maps

MODIS LST data were initially used for generating mean monthly climatology LST maps for June in the period of 2003-2008. The mean and maximum Aqua day and night LST values for June are presented in Fig. 10 for the period 2003-2008 for two urban areas (Nicosia, Larnaca) and one rural area (Ag. Marina). The curves show that the mean night LST values for the two urban areas are similar, while for the area of Ag. Marina, the temperature levels are 3-4 °C lower. For all sites, a minimum was observed for year 2005. The situation is different though regarding day LST values. The coastal site of Larnaca exhibited the lowest values among the three areas, while Nicosia and Ag. Marina exhibited similar patterns and temperature levels. The overall trend over time for the three areas showed a positive trend.

The intense heat wave event of 24 June 2007 was next examined in order to study the LST behavior during such events since satellite derived LST is controlled by land cover and topographic effect factors. In Fig.11, temperature anomaly maps, in terms of temperature deviation from the long-term monthly mean values (calculated for the period 2003-2008), are presented for the heat wave event under consideration and for both night and day Aqua passes.

The spatial patterns observed in the temperature anomaly maps are complex. It can be observed that day LST anomaly is more intense (up to 15°C) than the night anomaly. Minimum anomaly is located in the area of the mountain range Troodos (central-eastern

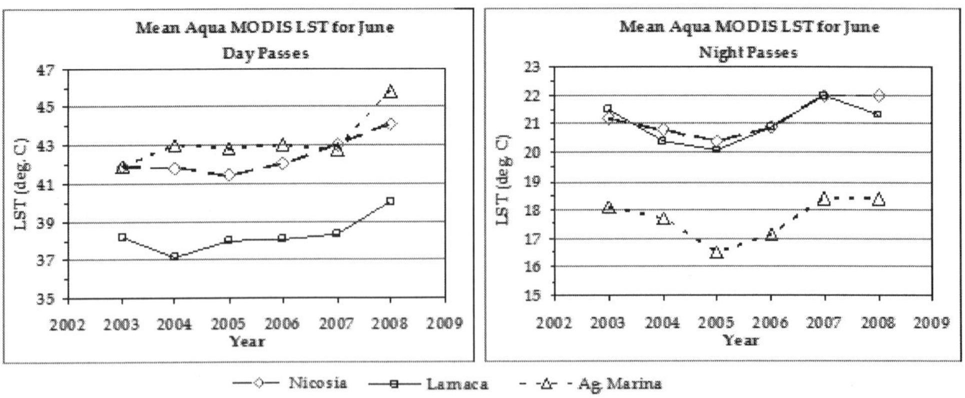

Figure 10. Yearly evolution of the mean Aqua MODIS LST for June (2003-2008) as retrieved from Aqua satellite for three different areas in Cyprus

part), while the southern part of Cyprus presents higher anomaly values than the northern part. The different values of LST increase are attributed to the difference in the emitted radiance from each land type and/or the urban heat island effect.

The amplitude of LST anomaly variation between day and night was examined with the land cover types based on the CORINE 2000 land cover map (Fig. 12). It was found that the mean anomaly amplitude was 2.89-4.05°C for artificial surfaces, 2.87-6.01°C for agricultural areas and 2.81-4.63°C for forest and semi natural areas. However, variations were noticed even in the same category. For example, for artificial surfaces the higher amplitude was noticed for airports and the lower for dump sites. For agricultural areas,

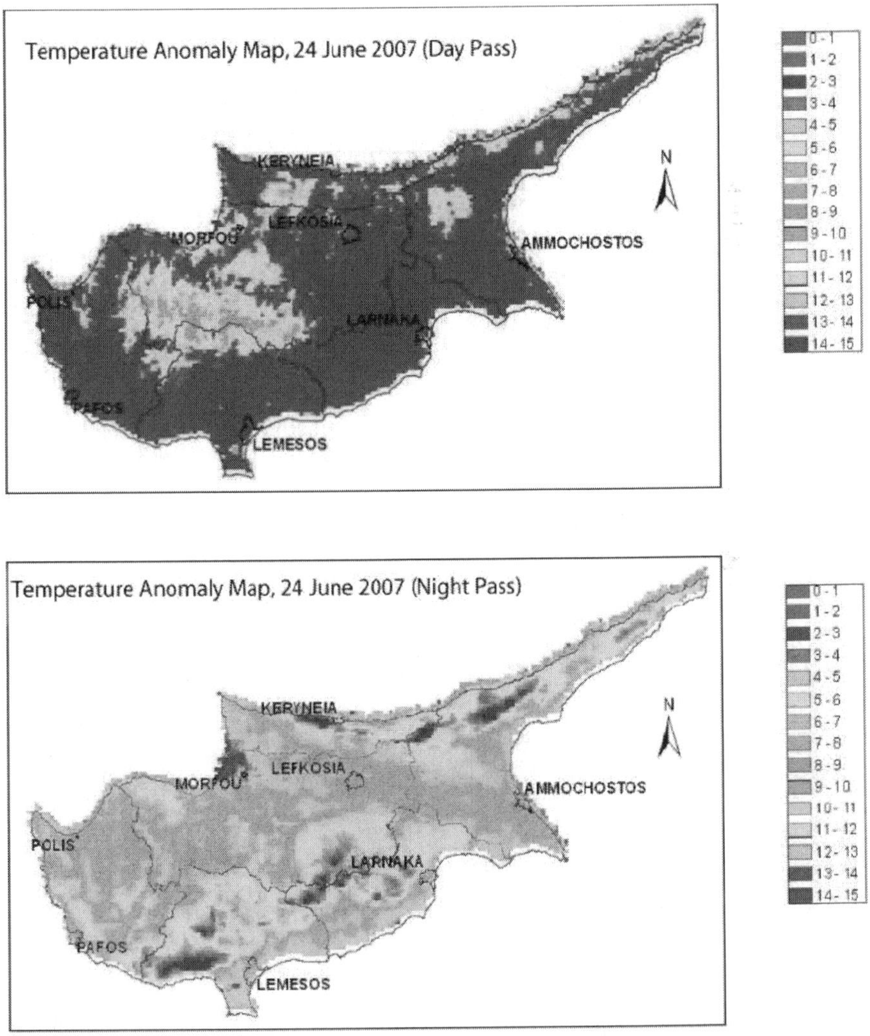

Figure 11. Land Surface Temperature anomaly map derived from both day (top) and night (bottom) Aqua MODIS passes for the selected heat wave event

the higher amplitude was noticed for pastures and the lower for annual crops associated with permanent crops. For forest and semi natural areas, the higher amplitude was noticed for beaches, dunes and sands and the lower for mixed forest.

A close inspection on the Aqua LST image (Fig. 12) acquired on 24 June 2007 (day pass) depicted that the highest LST values are noticed in areas that are recognized as vulnerable to desertification (Fig. 13). In Cyprus, there are two climatic zones that are considered as sensi- tive to desertification: the semi-arid, which extends over the larger part of the island and the arid sub-humid, which covers the slopes of the Troodos range and the windward side and higher parts of the Kyrenia range (IACO, 2007).

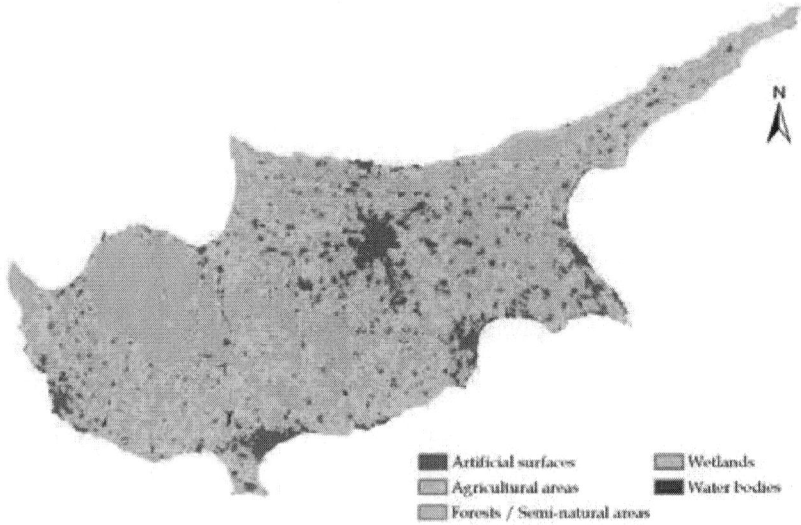

Figure 12. Simplified CORINE 2000 Land Cover map of Cyprus

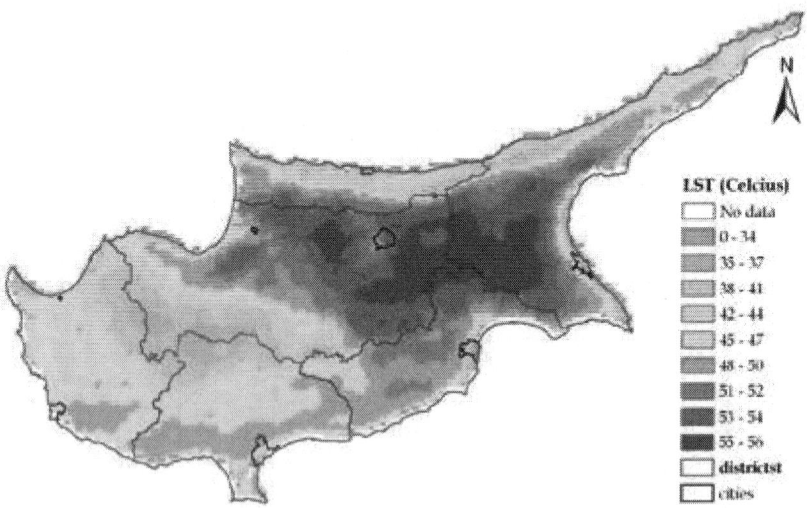

Figure 13. Land Surface Temperature map derived from day Aqua pass for the selected heat wave event

6. Urban heat island analysis

The variation of the UHI magnitude was examined for the four urban areas of Cyprus based on MODIS Aqua images acquired at night-time (at approximately 0130 local time). The selection of the MODIS Aqua data was based on the criterion that the night-time acquired images allow a more precise LST calculation since there is no incoming solar radiation to change the surface radiation balance, while night-time MODIS LST accuracy has been found to be better than day time (Rigo et al., 2006).

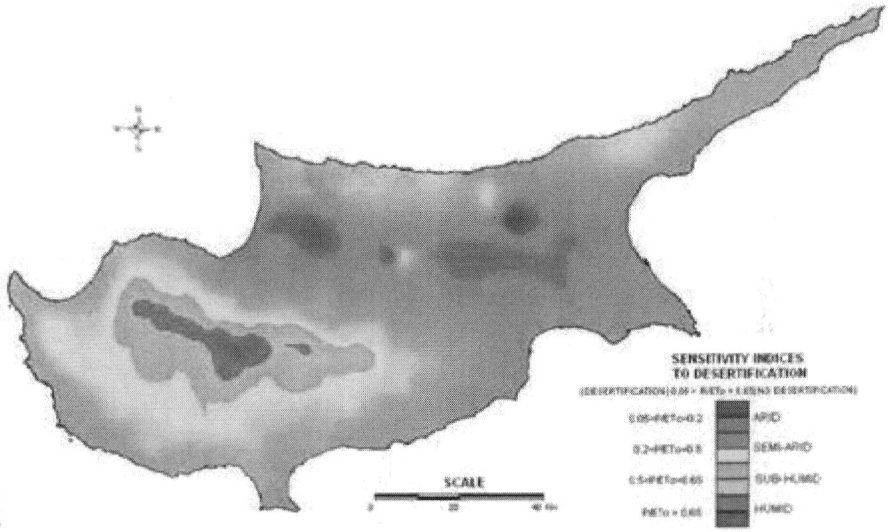

Figure 14. Areas sensitive to desertification, according to the United Nations Convention to Combat Desertification (IACO, 2007)

The magnitude of the UHI was estimated for each of the four test sites both for the mean monthly period 2002-2008 (Fig. 14) and for selective days of high temperature records of Au- gust 2010 (Fig 15). The UHI magnitude was calculated by subtracting the LST value from a rural area (as identified from the position of a pre-selected rural meteorological station) from the respective LST values falling within the urban boundary area of each district on a pixel-by-pixel basis (Tomlinson et al., 2010).

Fig.14 presents the mean monthly maximum UHI intensity for the period 2002-2008 for the four urban areas of Cyprus. As noticed, Nicosia, which is located in the centre of Cyprus, is most vulnerable to UHI during the warm period, when the intensity is recorded above four degrees. On the contrary, the other urban areas (Larnaca, Limassol and Paphos), which are close to the coastline, are lesser affected by UHI during the warm period, with intensities recorded around 1.5 to 3.5°C. These areas also demonstrated high UHI intensities during the cold period.

Next, the spatio-temporal variation of the UHI intensity for each of the urban areas was examined for the period 23 July to 28 August 2010, when high air surface temperatures were recorded (Fig. 14). The temporal variation of the maximum UHI intensity was estimated from the available nocturnal Aqua MODIS images for that period. The results revealed that, for most of the cases, the UHI magnitude curves follow a similar trend. Two major peaks were observed, on 31 July and 25 August 2010.

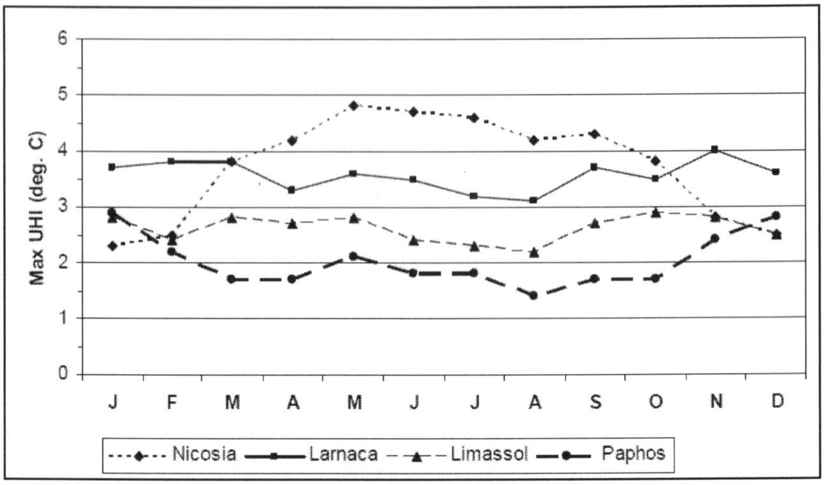

Figure 15. Mean monthly maximum UHI magnitude estimated from MODIS Aqua nocturnal images for the period 2002-2008 for Nicosia, Larnaca, Limassol and Paphos

The spatial variation of the UHI magnitude (Fig. 16) was examined for two dates (31 July and 28 August 2010) and was compared to the mean UHI magnitude as calculated for August for the years 2002-2008. The results derived suggest that, in almost all cases, the spatial patterns of the UHI magnitude observed for each urban area are quite simi- lar to each other with a few variations in the magnitude of intensity due to the severi- ty of the heat wave event. The highest intensities were noticed within the areas of the urban fabric.

The maximum intensities of UHI for each urban area were (a) 31 July 2010: 5.2°C (Nicosia), 3.5°C (Larnaca), 1.9°C (Limassol), and 5.0°C (Paphos) and (b) 25 August 2010: 6.9°C (Nico- sia), 3.9°C (Larnaca), 3.1°C (Limassol), and 4.2°C (Paphos). Thus, the deviation form the mean monthly UHI intensities calculated for July and August, correspondingly, were of about 0.6°C and 2.7°C for Nicosia, 0.3°C and 0.8°C for Larnaca, -0.4°C and 1.9°C for Limas- sol and 3.2°C and 2.8°C for Paphos.

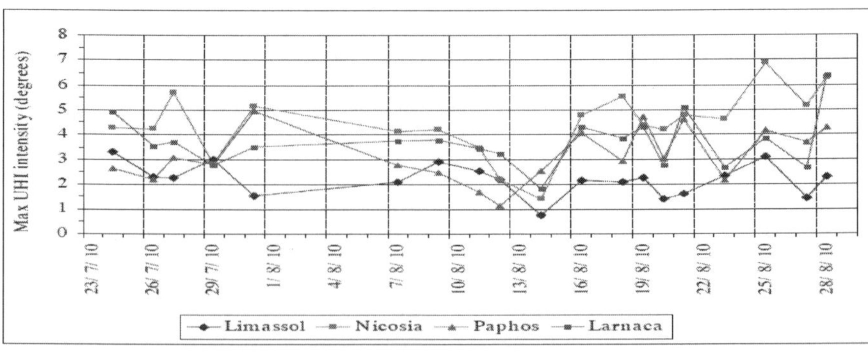

Figure 16. Temporal variation of maximum UHI intensity for the four urban areas of Cyprus, as derived from the analy- sis of Aqua nocturnal data for the period 23 July to 28 August 2010

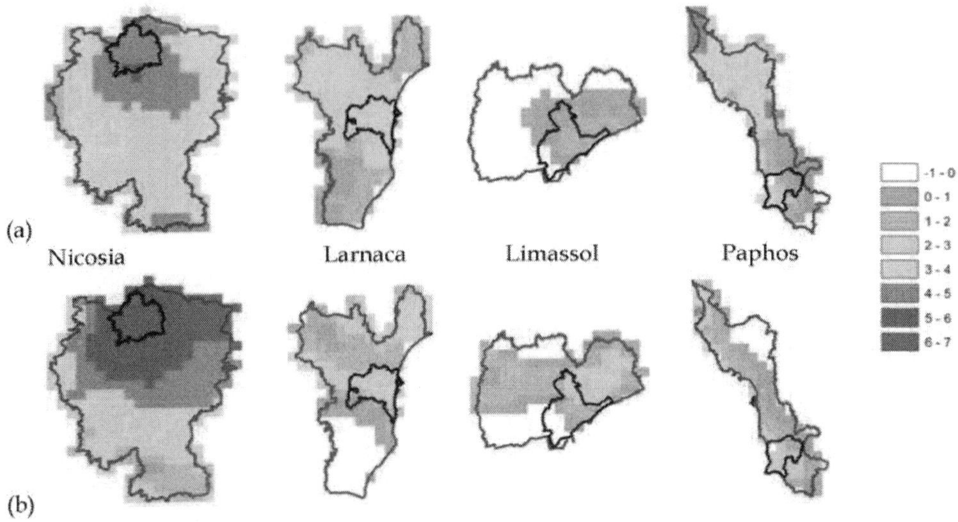

Figure 17. UHI estimated from MODIS Aqua nocturnal images for (a) 31 July and (b) 28 August 2010, for the four urban areas of Cyprus, separately

7. Concluding remarks

This Chapter commented on the use of Earth observation data along with ground meteorological data for the study of the UHI phenomenon in Cyprus. The synoptic conditions favoring the development of heat wave events were discussed. Neural Network analysis was used for classifying synoptic patterns and relate them with heat events. The majority of the heat events have occurred during the transition periods (Spring and Autumn). However, despite the fact that some clusters can be associated with such phenomena the connection between these events and atmospheric circulation at 500hPa did not give clear results.

Furthermore, an attempt was made in order to correlate ground temperature measurements and MODIS LST data. The results have shown that the methodology can perform sufficiently; however, further refinement is needed in order to improve this approach.

Aqua MODIS retrievals of land surface temperature data were used for studying selective heat wave events. The analysis of LST data depicted the regions that are more prone to such events. The spatial variations of the UHI magnitude was also examined for the major cities of Cyprus, during both mean monthly conditions and for selected events, identifying areas that are most vulnerable.

References

1. Cannon, J.A. & Whitfield, P.H. (2002). Synoptic map-pattern classification using recursive partitioning and principal component analysis, *Monthly Weather Review*, Vol. 130, pp. 1187–1206
2. Coll C.; Caselles V.; Galve J.; Valor E.; Niclos R. ; Sanchez J. & Rivas R. (2005). Ground measurements for the validation of land surface temperatures derived from AATSR and MODIS data, *Remote Sensing of Environment*, Vol.97, pp. 288–300, DOI: 10.1016/j.rse.2005.05.007

3. Coll, C.; Wan, Z. & Galve J.M. (2009). Temperature-based and radiance-based validations of the V5 MODIS land surface temperature product, *Journal of Geophysical Re- search*, 114, D20102, doi:10.1029/2009JD012038

4. Dousset, B. & Gourmelon, F. (2003). Satellite multi-sensor data analysis of urban surface temperatures and landcover, *ISPRS,* 58, pp. 43–54

5. El-Kadi, A.K.A. & Smithoson, P.A. (1992). Atmospheric classifications and synoptic climatology, *Progress in Physical Geography*, 16, pp.432–455

6. Founda, D. & Giannakopoulos, C. (2009). The exceptionally hot summer of 2007 in Athens, Greece: A typical summer in the future climate, *Global and Planetary Change,* 67, pp.227-236, doi:10.1016/j.gloplacha.2009.03.013

7. Hadjimitsis D. & Agapiou A. (2011). A remote sensing evaluation of urban expansion and its impact on Urban Heat Island phenomenon: the case study of Limassol, Cy- prus, *Proceedings of the 5th International Conference Earth From Space - the Most Effective Solutions*, 29/11 - 01/12/2011, Moscow, Russia, p.301-303, ISBN:978-5-9518-0490-7

8. Haykin, S. (1998). *Neural Networks: A Comprehensive Foundation,* 2nd edition, Macmillan College Publishing

9. Hewitson, B.C. & Crane, R.G. (1996). Climate downscaling: Techniques and application. *Climate Research*, 7, pp.85–95

10. Hulley, G.C. & Hook, S.J. (2009). Intercomparison of versions 4, 4.1 and 5 of the MODIS Land Surface Temperature and Emissivity products and validation with lab- oratory measurements of sand samples from the Namib desert, Namibia, *Remote Sensing of Environment*, 113, pp. 1313-1318

11. Hung, T.; Uchihama, D.; Ochi, S. & Yasuoka, Y. (2006).Assessment with satellite data of the urban heat island effects in Asian mega cities, *International Journal of Applied Earth Observation and Geoinformation,* 8, pp. 34–48

12. IACO (2007). *Consultation Services for the Production of a National Action Plan to Combat Desertification in Cyprus, I,* IACO Environmental and Water Consultants Ltd, Cyprus.

13. Imhoff, M. L., Zhang, P., Wolfe, R. E., Bounoua, L. (2010). Remote sensing of the ur- ban heat island effect across biomes in the continental USA, *Remote Sensing of Envi- ronnent,* 114 (3), pp. 504–513

14. Kalogirou, S.; Michaelides, S. & Tymvios, F. (2002). Prediction of maximum solar ra- diation using artificial neural networks. *World Renewable Energy Congress VII (WREC 2002),* Cologne, 29 June – 5 July, 2002, A.A. Sayigh (Ed.), Elsevier Science Ltd (on CD- ROM)

15. Kato, A. & Yamaguchi, Y. (2005). Analysis of urban heat-island effect using ASTER and ETM+ Data: separation of anthropogenic heat discharge and natural heat radiation from sensible heat flux, *Remote Sensing of Environment,* 99, pp. 44–54

16. Key, J. & Crane, R. J. (1986). A Comparison of Synoptic classification schemes based on “objective” procedures, *Journal of Climatology*, 6, pp. 375–388

17. Kohonen, T. (1990). The Self-Organizing Map, *Proceedings of the IEEE*, 78, pp.1464–1480
18. Lo, C.P. & Quattrochi, D.A., (2003). Land-use and land-cover change, urban heat island phenomenon, and health implications: a remote sensing approach, *Photogrammetric Engineering & Remote Sensing*, Vol.69, pp. 1053–1063
19. Monteiro, I.; Trigo, I.F.; Kebasch, E. & Olesen F. (2007). Validation of land surface temperature retrieved from Meteosat Second Generation Satellites, *Proceedings of the Joint 2007 EUMETSAT Meteorological Satellite Conference and the 15th Satellite Meteorol- ogy & Oceanography Conference of the American Meteorological Society*, Amsterdam, The Netherlands, 24-28 September 2007, Available from: http://www.eumetsat.int/Home/ Main/Publications/index.htm
20. Mostovoy, G.V.; King, R.; Reddy, K.R. & Kakani, V.G. (2005). Using MODIS LST data for high-resolution estimates of daily air temperature over Mississippi, *Proceedings of the International Workshop on the Analysis of Multi-Temporal Remote Sensing Images*, 16-18 May 2005, pp. 76-80
21. Michaelides, S.C. ; Evripidou, P. & Kallos, G. (1999). Monitoring and predicting Saharan desert dust transport in the eastern Mediterranean, *Weather*, Vol.54, pp.359–365
22. Michaelides, S. ; Tymvios , F. & Michaelidou, T. (2009). Spatial and temporal characteristics of the annual rainfall frequency distribution in Cyprus, *Atmospheric Research*, Vol.94, pp. 606–615
23. Michaelides, S.; Tymvios, F.S. & Charalambous, D. (2010). Investigation of trends in synoptic patterns over Europe with artificial neural networks, *Advances in Geosciences*, Vol.23, pp.107-112
24. Mihalakakou, G.; Flocas, H.A. ; Santamouris, M. & Helmis, C.G. (2002). Application of Neural Networks to the Simulation of the Heat Island over Athens, Greece, Using Synoptic Types as a Predictor, *Journal of Applied Meteorology*, Vol.41, pp. 519-527
25. Nichol, J.E. (1996). High resolution surface temperature patterns related to urban morphology in a tropical city: a satellite-based study, *Journal of Applied Meteorology*, Vol.35, pp. 135–146
26. Nichol, J. E.; Fung, W. Y.; Lam, K. & Wong, M. S. (2009). Urban heat island diagnosis using ASTER satellite images and 'in situ' air temperature, *Atmospheric Research*, Vol. 94, pp. 276–284
27. Peng, S.; Piao, S.; Ciais, P.; Friedlingstein, P.; Ottle, C.; Bréon, F-M.; Nan, H.; Zhou, L. & Myneni, R. B. (2012). Surface Urban Heat Island across 419 global big cities, *Environmental Science & Technology*, Vol.46, pp. 696-703, dx.doi.org/10.1021/es2030438
28. Retalis, A.; Paronis, D. ; Lagouvardos, K. & Kotroni, V. (2010). The heat wave of June 2007 in Athens, Greece—Part 1: Study of satellite derived land surface temperature, *Atmospheric Research*, Vol.98, pp. 458–467
29. Rigo, G.; Parlow, E. & Oesch D. (2006). Validation of satellite observed thermal emission with in-situ measurements over an urban surface, *Remote Sensing of Environment*, Vol.104, pp. 201–210, doi: 10.1016/j.rse.2006.04.018

30. Snyder, W. C.; Wan, Z.; Zhang, Y. & Feng, Y.-Z. (1998). Classification-based emissivity for land surface temperature measurement from space. *International Journal of Re- mote Sensing,* Vol.19, pp. 2753–2774

31. Streutker, D.R. (2002). A remote sensing study of the urban heat island of Houston, Texas, *International Journal of Remote Sensing,* Vol.23, pp. 2595–2608

32. Tomlinson, C.; Chapman, L.; Thornes J. & Baker C. (2010) Derivation of Birmingham's summer surface urban heat island from MODIS satellite images, *International Journal of Climatology*, doi:10.1002/joc.2261

33. Tran, H.; Uchihama, D.; Ochi, S. & Yasuoka, Y. (2006). Assessment with satellite data of the urban heat island effects in Asian mega cities, *International Journal of Applied Earth Observation and. Geoinformation,* Vol.8, pp. 34–48

34. Tymvios, F.; Jacovides, K.P. & Michaelides, S.C. (2002). Calculation of global solar ra- diation on a horizontal surface by using Artificial Neural Networks, *Proceedings of the 6th Panhellenic Conference on Meteorology, Climatology and Atmospheric Physics,* Ioanni- na, Greece, 25-28 September, 2002 (In Greek) pp. 468-475

35. Tymvios, F.S.; Jacovides, C.P. ; Michaelides, S.C. & Skouteli, C. (2005a). A comparative study of Ångstrom's and artificial neural networks' methodologies in estimating global solar radiation, *Solar Energy*, Vol.78, pp.752-762

36. Tymvios, F.; Jacovides, C. ; Michaelides, S. ; Schizas, C. & Scouteli, C. (2005b). Modelling the photosynthetically active radiation (PAR) for Nicosia in Cyprus, *Proceedings of the 7th Panhellenic (International) Conference of Meteorology, Climatology and Atmos- pheric Physics*, 28-30 Sept., 2004, Nicosia, Cyprus, pp. 972-980

37. Tymvios, F.S. ; Constantinides, P. ; Retalis, A. ; Michaelides, S. ; Paronis, D. ; Evripidou, P. & Kleanthous S. (2007). The AERAS project– database implementation and Neural Network classification tests, *Proceedings of the 6th International Conference on Urban Air Quality*, Limassol, Cyprus, 27-29 March 2007

38. Tymvios F.S.; Savvidou K.; Michaelides S.C. & Nicolaides K.A. (2008). Atmospheric circulation patterns associated with heavy precipitation over Cyprus, *Geophysical Re- search Abstracts*, Vol.10, EGU2008-A-04720

39. Tymvios, F.; Savvidou, K. & Michaelides, S. (2010a). Association of geopotential height patterns with heavy rainfall events in Cyprus, *Advances in Geosciences*, Vol.23, pp.73-78

40. Tymvios F.; Charalambous D ; Michaelides S. ; Retalis A. ; Paronis D. & Skouteli C. (2010b). Temperature distribution in Cyprus with the use of satellite images and arti- ficial neural networks, *Proceeding of the 10th International Conference on Meteorology, Climatology and Atmospheric Physics*, 25-28 May 2010, Patra, Greece, 227-234

41. Wan Z. & Dozier J. (1996). A generalized split-window algorithm for retrieving land-surface temperature from space, *IEEE Transactions on Geoscience and Remote Sensing,* Vol.34, pp. 892–905, doi:10.1109/36.508406

42. Wan Z. (1999). *MODIS Land-Surface Temperature Algorithm Theoretical Basis Document (LST ATBD),* University of California, Santa Barbara, USA, Institute for

Computa- tional Earth System Science, Available from: http://modis.gsfc.nasa.gov/data/atbd/ atbd mod11.pdf

43. Wan Z. (2002). Validation of the land-surface temperature products retrieved from Terra Moderate Resolution Imaging Spectroradiometer data, *Remote Sensing of Environment,* Vol.83, pp. 163–180, doi:10.1016/S0034-4257(02)00093-7

44. Wan Z. ; Zhang Y. ; Zhang Q. & Li Z. (2004). Quality assessment and validation of the MODIS global land surface temperature, *International Journal of Remote Sensing,* Vol.25, pp. 261–274, doi:10.1080/0143116031000116417

45. Wan, Z. (2008). New refinements and validation of the MODIS land-surface temperature/emissivity products, *Remote Sensing of Environment*, Vol.112, pp.59-74

46. Xiao, R.; Ouyang, Z.; Heng, H.; Li, W.; Schienke, E. & Wang, X. (2007). Spatial pattern of impervious surfaces and their impacts on land surface temperature in Beijing, China. *Journal of Environmental Science* , Vol.19, pp. 250–256

47. Xoplaki, E.; Gonzalez-Rouco, J.F.; Luterbacher, J. & Wanner, H. (2003). Mediterra- nean summer air temperature variability and its connection to the large-scale atmos- pheric circulation and SST, Climate Dynamics, Vol.20, pp.723-739

48. Yuanbo Liu, Y.; Yamaguchi, Y. & Ke, C. (2007). Reducing the discrepancy between ASTER and MODIS Land Surface Temperature products, *Sensors*, Vol.7, pp. 3043–3057

Chapter-2

Remote Sensing for Determining Evapotranspiration and Irrigation Demand for Annual Crops

1. Introduction

Evapotranspiration (ETc) is the mean for exploiting irrigation water and constitutes a major component of the hydrological cycle (Telis et al., 2007; Papadavid, 2011). The ETc is a basic and crucial parameter for climate studies, weather forecasts and weather modeling, hydro- logical surveys, ecological monitoring and water resource management (Hoedjes et al., 2008). In the past decades, the estimation of ETc combining conventional meteorological ground measurements with remotely-sensed data, has been widely studied and several methods have been developed for this purpose (Tsouni, 2003). For hydrological resources management and irrigation scheduling, an accurate estimation of the ETc is necessary to be considered (Hoedjes et al., 2008 ; Papadavid et al., 2011). Crop evapotranspiration rate is highly important in various areas of the agricultural sector such as for identification of crop stress, water deficiency, for estimating the exact potential needs of crops for best yields. It is well accepted that water depletion methods, such as lysimeters, are the most accurate methods for estimating ETc. Methods that use meteorological parameters in order to estimate the ETc of different crops are well established and used by various studies (Telis et al., 2007; Rogers et al., 2007). A number of semi-empirical methods have been also developed in order to estimate the evapotranspiration from different climatic variables (Courault et al., 2005). Remotely sensed reflectance values can be used in combination with other detailed information for estimating ETc of different crops. Indeed, the potentiality of remote sensing techniques in ETc estimation and water resource management has been widely acknowledged (Papadavid et al., 2010). The possibility for monitoring irrigation demand from space is an important factor and tool for policy makers. It has been found that saving irrigation water through remote sensing techniques could diminish farm irrigation cost which reaches 25% of the total costs and increases the margin of net profit (Papadavid et al., 2011). Several re- searchers such as D'Urso et al., (1992), Bastiaanssen (2000), Ambast et al., (2006) and Papada- vid et al., (2011) have highlighted the potentiality of multispectral satellite images for the appraisal of irrigation management. The integration of remotely sensed data with auxiliary ground truth data for obtaining better results is common in the literature. (Bastiaanssen et al., 2003; Ambast et al., 2006; Minaccapili et al., 2008). Ambast et al., (2006) have shown that the application of remote sensing data in irrigation is of high importance because it supports management of irrigation and is a powerful tool in the hands of policy makers. It has been found that research in ETc is directed towards energy balance algorithms that use remote sensing directly to calculate input parameters and, by combining empirical relationships to physical models, to estimate the energy budget components (Minaccapili et al., 2008; Papa- david et al., 2010; Papadavid et al., 2011). All the remote sensing models of this category are characterized by several approximations and need detailed experimental validations. Multi- spectral images are used to infer ETc, which is the main input for water balance methods- models. For estimations of ET, ground truth data (Leaf Area Index-LAI, crop height) and meteorological data (air temperature, wind speed, humidity) is needed to support this ap- proach. In nearly every application of water balance model, knowledge of spatial variations in meteorological conditions is needed (Moran et al., 1997).

The use of remote sensed data is very useful for the deployment of water strategies since it can offer a huge amount of information in short time, compared to conventional methods. Besides convenience and time reducing, remotely sensed data lessens the costs for data acquisition, especially when the area is extended (Thiruvengadachari et al., 1997). Although remote sensing based ETc models have been shown to have the potential to accurately esti- mate regional ETc, there are opportunities to further improve these models testing the equa- tions used to estimate LAI and crop height for their accuracy under current agro- meteorological and soil conditions.

This Chapter discusses the implementation of the most widely used models for estimating ETc, the 'SEBAL' and 'Penman-Monteith' which are used with satellite data. Such models are employed and modified, with semi-emprical models regarding crop canopy factors, to estimate accurately ETc for specific crops in the Cyprus area under local conditions. Crop Water Requirements have been determined based on the evapotranspiration values.

2. Study area

The study area is located in the area of Mandria village, in the vicinity of Paphos International Airport in Paphos District in Cyprus (Figure 1). The study area lying in the southwest of Cyprus is a coastal strip between Kouklia and Yeroskipou villages. The area is a coastal plain with seaward slope of about 2% and it consists of deep fertile soils made of old fine deposits. The area is dissected by three major rivers, the Ezousa, Xeropotamos and Diarizos. The area is almost at sea level (altitude 15 m) and is characterized by mild climate which provides the opportunity for early production of leafy and annual crops. The uniform and moderate temperatures, attributed to the permanent sea breeze of the area, and the relative humidity, are conductive to the early production of fruits and vegetables, for which the rep- utation of the area is known all over Cyprus. Cereals are also cultivated in the area. A typi- cal Mediterranean climate prevails in the area of interest, with hot dry summers from June to September and cool winters from December to March, during which much of the annual rainfall occurs with an average record of 425mm. Nevertheless, irrigation is indispensable for any appreciable agricultural development in the area

The selected area is a traditionally agricultural area with a diversity of annual and perennial cultivations and is irrigated by Asprokremnos Dam, one of the biggest dams of Cyprus.

Figure 1. Partial Landsat TM image of Mandria Village in the vicinity of Paphos International Airport in Cyprus

3. Resources

3.1 Field spectroradiometer

The GER (Geophysical Environmental Research) 1500 field spectroradiometer (Figure 2,3) is a light-weight, high performance, single-beam field spectroradiometer. It is a field portable spectroradiometer covering the ultraviolet, visible and near-infrared wavelengths from 350 nm to 1050 nm. It uses a diffraction grating with a silicon diode array which has 512 discrete detectors and provides the capability to read 512 spectral bands.

Figure 2. Spectroradiometric measurement over spectralon panel (Papadavid, 2012)

The instrument is very rapidly scanning, acquiring spectra in milliseconds. The spectroradi- ometer provides the option for stand-alone operation (single beam hand-held operation) and the capability for computer assisted operation through its serial port, which offers near real-time spectrum display and hard disk data transfer. The maximum number of scans (512 readings), can be stored for subsequent analysis, using a personal computer and GER li- censed operating software. The Lens barrel used for the specific spectroradiometer is the Standard 4 field of view. The data are stored in ASCII format for transfer to other software.

Reflectance factors using a control stable surface with known characteristics as described by McCloy (1995) have been measured. Many researchers (McCloy, 1995; Beisl, 2001; Anderson and Milton, 2006; Schaepman, 2007; Papadavid 2012) highlight the advantages of using con- trol surfaces in the measurement of reflectance factors (Bruegge et al., 2001). In this study, the control surface was a commercially available "Labsphere" compressed "Spectralon" white panel (Figure 2). There is evidence that these types of panels are more consistent and retain their calibration better than painted panels (Jackson et al., 1992; Beisl, 2001). Spectra- lon diffuse reflectance targets are ideal for laboratory and field applications such as field val- idation experiments, performed to collect remote sensing data due to the fact are: durable and washable; have typical reflectance values of 95% to 99% and are spectrally flat over the UV-VIS-NIR spectrum; are impervious to harsh environmental conditions and chemically inactive (Papadavid et al., 2011). Reflectance was calculated as a ratio of the target radiance to the reference radiance. The target radiance value is the measured value taken on the crops (Figure 3) and the reference radiance value is the measured value taken on the standard Spectralon panel (Figure 2), representing the sun radiance, which reaches the earth surface—without atmospheric influence.

Figure 3. Spectroradiometric measurements over potatoes (target) in Mandria Village in Paphos, Cyprus (Papada- vid, 2012)

3.2 SunScan canopy analyser system

Leaf Area Index is commonly used for monitoring crop growth. Instead of the traditional, direct and labor-consuming method of physically measuring the plant with a ruler (direct method), an optical instrument, SunScan canopy analyser system (Delta-T Devices Ltd., UK) is used (indirect method). The instrument (Figure 4) is indirectly measuring LAI by measuring the ratio of transmitted radiation through canopy to incident radiation (Figure 5). Indirect methods for LAI measurements based on the transmittance of ra- diation through the vegetation have been developed (Lang et al., 1991; Welles and Norman, 1991).

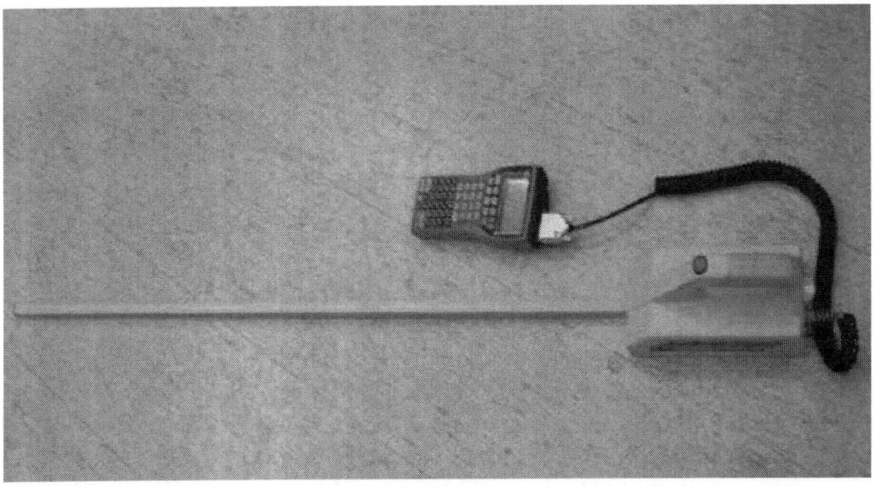

Figure 4. SunScan (Delta-T) canopy analyser for LAI and crop height measurements

Figure 5. Use of SunScan canopy analyzer for LAI measurements (Papadavid, 2012)

3.3 Satellite imagery

Spatial, spectral and temporal resolution of satellite images is very important for studies dealing with crop water requirements. Landsat- 5 TM and -7 ETM+ have been widely used for hydrological studies due to their relatively good temporal resolution (16 days) which is important for providing regular snapshots during the crop growth season (Dadhwal et al., 1996; Song et al., 2001; Alexandridis, 2003). These sensors are suitable for agricultural areas with medium to big average fields due to their high spatial resolution (60m for thermal band, 15m for panchromatic and 30m for the rest).

However, the availability of images is highly dependent on weather conditions. The availability of cloud free images for operational projects is very important and depends on the geographical position and the prevailing weather conditions for the area of interest (Kontoes and Stakenborg, 1990; Hadjimitsis et al., 2000; Hadjimitsis et al., 2010). Countries such as Greece and Cyprus are characterised by good weather conditions and the availability of cloud-free images (Hadjimitsis et al., 2000). An advantage of Landsat image for applications in Cyprus is that of the whole island coverage from a single image which can be inferred on a regular basis since Landsat satellites overpass Cyprus on a systematic basis (Papadavid, 2011). Remote sensing users or policy makers or governmental officers have the oppotunity to have remotely sensed data very often which is very important in terms of phenological cycle monitoring.

D' Urso (1995) and Minacapilly et al., (2008) explored the importance of using image time series due to the high importance of water requirements in the different stages of the crops. The same crop in different stage has different water needs, therefore the time ser- ies of satellite images is very important in studies regarding ETc and remote sensing. A time series of Landsat 5 TM and 7 ETM+ imagery acquired in years 2008, 2009 and 2010 are used in this study, as listed in Table 1. The crucial aim is to have satellite images in all stages of the specific crops. The availability of images is important since these images will be converted into ETc maps using an image processing software such as ERDAS Imagine software. Hence, the more images we have the better analysis we get. All im- ages were pre-processed in order to remove atmospheric and radiometric effects, using the ERDAS Imagine software. ERDAS 'modeller module' was used to transform the im- ages into maps by applying the ETc algorithms. The same satellite images were also used for evaluating the statistical models found, regarding Leaf Area Index (LAI) or Crop Height (CH) to one of the selected Vegetation Indices (VI).

4. Methodology

An attempt has been made to statistically describe the crop canopy factors, namely crop height (CH) and LAI, through the vegetation indices (VI). Crop canopy factors are vital elements in the procedure of estimating ETc. These indices were produced from spectroradiometric measurements using a hand-held field spectroradiometer (GER 1500) and after this data were filtered through the Relative Spectral Response (RSR) filters of the corresponding Landsat TM/ETM+ bands. At the same time LAI and CH direct measurements were taken in situ. Hence, time series of LAI, CH and VI have been created and were used to model LAI and CH to VI. After applying the needed regression analysis and evaluating them, the best model for each crop, based on the determination factor (r^2), was used in specific ETc algo- rithms in a procedure to adapt and modify the algorithms in the geo-morphological and meteorological conditions of the island of Cyprus as a representative Mediterranean region.

Table 1. Landsat TM/ETM+ images used in this study (Papadavid, 2012)

	Satellite	Sensor	Date	DOI	Cos Zenith Angle	Potatoes	Ground Nuts	Beans	Chickpeas
1	Landsat-7	ETM+	12/7/2008	193	0,76		x	x	
2	Landsat-7	ETM+	28/7/2008	209	0,81		x	x	
3	Landsat-7	ETM+	13/8/2008	225	0,86		x	x	
4	Landsat-7	ETM+	29/8/2008	241	0,91		x	x	
5	Landsat-7	ETM+	14/9/2008	257	0,92	x	x	x	x
6	Landsat-7	ETM+	30/9/2008	272	0,83	x			x
7	Landsat-7	ETM+	16/10/2008	289	0,76	x			x
8	Landsat-7	ETM+	1/11/2008	305	0,66	x			x
9	Landsat-7	ETM+	17/11/2008	321	0,59	x			x
10	Landsat-5	TM	2/1/2009	2	0,79	x			x
11	Landsat-7	ETM+	21/2/2009	52	0,62	x			x
12	Landsat-5	TM	29/6/2009	179	0,91		x	x	
13	Landsat-5	ETM+	7/7/2009	188	0,91		x	x	
14	Landsat-5	TM	15/7/2009	196	0,90		x	x	
15	Landsat-5	ETM+	23/7/2009	204	0,89		x	x	
16	Landsat-5	TM	16/8/2009	228	0,76		x	x	
17	Landsat-7	ETM+	13/4/2010	103	0,47	x			
18	Landsat-7	ETM+	31/5/2010	151	0,67	x	x		
19	Landsat-7	ETM+	16/6/2010	167	0,74	x	x		
20	Landsat-7	ETM+	24/6/2010	175	0,80	x	x		
21	Landsat-7	ETM+	10/7/2010	191	0,84		x	x	
22	Landsat-7	ETM+	27/8/2010	239	0,92			x	
23	Landsat-5	TM	7/11/2010	311	0,93			x	x
24	Landsat-7	ETM+	9/12/2010	343	0,65				x
25	Landsat-7	ETM+	2/5/2011	122	0,62	X		X	
26	Landsat-7	ETM+	19/6/2011	170	0,73	X	X	X	
27	Landsat-7	ETM+	5/7/2011	186	0,83		X		
28	Landsat-7	ETM+	21/7/2011	202	0,84		X		X
29	Landsat-5	TM	29/7/2011	210	0,86		X		X
30	Landsat-5	TM	30/8/2011	242	0,93		X		X

Crop water requirements were inferred by applying the algorithms and it was tested to check if the specific modifications have assisted the algorithms to improve their precision when estimating ETc.

The overall methodology is described below. The intended purpose is to estimate ETc of specific crops in the area of interest using remote sensing techniques.

- Spectroradiometric measurements were undertaken for two years (2009-10) in order to collect spectral signatures of each crop included in the study. For each crop (Potatoes, Groundnuts, Beans and Chickpeas) the average spectral signature in each phenological stage was created based on the two cultivating periods (2009-2010). The purpose is to have the reflectance of each crop during their phenological stages after the data was fil- tered through the Relative Spectral Response filters.

- ❖ Leaf Area Index (LAI) and Crop Height (CH) measurements were also taken simultane- ously to spectroradiometric measurements and following the same phenological cycle of each crop for the corresponding cultivating periods. The purpose was to create time series of these two parameters to correlate them to Vegetation Indices (VI).
- ❖ Development of vegetation indices (VI). Time series of VI were created based on the re- flectance of each crop, in each phenological stage.
- ❖ Modeling VI to LAI and CH. Different models were tested in order to identify the best possible model which better describes LAI and CH through VI.
- ❖ Preprocessing of satellite images was applied. Geometric rectification, radiometric correc- tion including atmospheric correction of satellite data were applied before main process- ing of the data.
- ❖ Mapping LAI, CH and albedo was performed. The three crop canopy parameters were mapped using the ERDAS Imagine v.10 software. The satellite images were transformed into maps in order firstly to test in practice the models and secondly to be inserted as in- puts in ETc algorithms.
- ❖ Models verification. After inferring the best model describing LAI or CH using VI, an evaluation of this procedure was taking place. A priori knowledge of satellite over pass- ing over the area of interest has assisted the procedure of taking LAI and CH measure- ments in different plots and different cultivating period. These average measurements were compared to the predicted measurements arising from the models application found in the previous step.
- ❖ Application of ETc algorithms. Original and modified by previous equations, ETc algo- rithms have been applied to check, based on the reference method, if and how the models have boosted accuracy on estimating ETc for each crop.

5. Ground data

5.1 Spectral signatures of crops

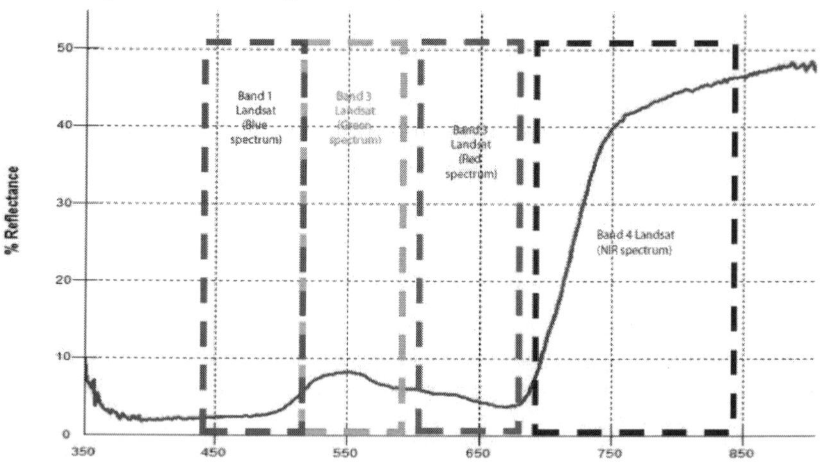

Figure 6. Vegetation spectral signature: Vegetation has low reflectance in the visible region and high reflectance in the near infrared (data analysis)

It is well established that the reflectance and transmission spectrum of leaves is a function of both the concentration of light absorbing compounds (chlorophylls, carotenoids, water, cel- lulose, lignin, starch, proteins, etc.) and the internal scattering of light that is not absorbed or absorbed less efficiently (Newnham and Burt, 2001; Dangel et al., 2003). Each crop has a dif- ferent spectral signature depending on the stage of its phenological cycle (Gouranga and Harsh, 2005; McCloy, 2010; Papadavid et al., 2011). A general view of the vegetation spectral signature is shown in Figure 6; there is strong absorption in blue and red part of the light spectrum while at green and infrared part there is light and strong reflectance, respectively.

The domain of optical observations extends from 400 nm in the visible region of the electro- magnetic spectrum to 2500 nm in the shortwave infrared region. The strong absorption of light by photosynthetic pigments dominates green leaf properties in the visible spectrum (400-700nm). Leaf chlorosis causes an increase in visible reflectance and transmission. The near-infrared region (NIR, 700-1100 nm) is a region where biochemical absorptions are limit- ed to the compounds typically found in dry leaves, primarily cellulose, lignin and other structural carbohydrates (Wang et al., 2005). However, foliar reflection in this region is also affected by multiple scattering of photons within the leaf, related to the internal structure, fraction of air spaces, and air-water interfaces that refract light within leaves. The reflectance and transmittance in the middle-infrared also termed the shortwave-infrared (SWIR, 1100 nm - 2500 nm) is also a region of strong absorption, primarily by water in green leaves (Mai- er, 2000). More specifically, visible blue and red are absorbed by the two main leaf pigments, chlorophyll a and b in green-leaf chloroplasts. These strong absorption bands induce a re- flectance peak in the visible green. Thus most vegetation has a green-leafy color. Chloro- phyll pigments are also known as the green pigments.

Apart from chlorophyll, other leaf pigments have a significant effect on the visible spectrum. Carotene, a yellow to orange-red pigment strongly absorbs radiation in the 350-500 nm range and is responsible for the color of some flowers and leaves without chlorophyll. Xan- thophyll, the red and blue pigment also strongly absorbs radiation in the 350-500 nm range, giving the distinctive color to the leaves in Autumn. In the near infrared range (700-1000 nm) of the electromagnetic spectrum, there is strong reflectance in the spongy mesophyll cells that occur at the back of leaves.

5.2. Phenology of the crops

Phenology can be defined as the study of the timing of biological events, the causes of their timing with regard to biotic and abiotic forces, and the interrelation among phases of the same or different species (Shaykewich 1994). As McCloy (2010) mentions the phenological cycle can be defined as the trace or record of the changes in a variable or attribute over the phenological period (usually one agricultural year) and a phenophase is defined as an observable stage or phase in the seasonal cycle of a plant that can be defined by start and end points. Crop phenological stages are important indicators in agricultural production, management, planning, decision-making and irrigation scheduling (O' Leary et al., 1985; Gouranga and Harsh, 2005; Papadavid et al., 2011). Indeed, Food and Agriculture Organization (FAO) guidelines of estimating crop evapotranspiration for irrigation demands, take into account crop characteristics and the phenological stages of a crop; Crop coefficient (K_c)refers to crop growth stage and the length in time of this stage (Allen et al., 1998). Moreover crop phenology is difficult to be studied for large areas using traditional techniques and methods.

Recently, many studies have been performed in order to derive the crop phenological stages based on satellite images (Papadavid, 2011). These studies aim to validate vegetation indices for monitoring the development of the phenological cycle from times series data (Papadavid, 2011). For example, Sakamoto et al., (2005), Minaccapili et al., (2008) and Papadavid et al., (2011) used times series of remotely sensed data in order to develop a new systematic method for detecting the phenological stages of different crops from satellite data while Bradley et al., (2007) in their study have introduced a curve fitting procedure in order to derive inter-annual phenologies from time series of noisy satellite NDVI data. Moreover, Funk and Budde (2009) have used an analogous metric of crop performance based on time series of NDVI satellite imagery. Papada- vid et al., (2009; 2010; 2011) and Papadavid (2011) have shown that field spectroscopy and empirical modelling, when successfully integrated, can develop new models of Leaf Area Index (LAI) and Crop Height, during the phenological cycle of crops.

Tables indicating the phenology of each crop can be found in the FAO internet site (www.fao.org). Table 2 indicates the phenological stages of each crop and the number of *in situ* measurements (spectroradiometric and LAI/CH) taken at each stage.

Table 2. Phenological stages of each crop (Papadavid, 2012)

Phenological Stages of *Potatoes*	**GER1500 measurements**	**SunScan measurements**	**Phenological Stages of *Peas***	**GER1500 mean measurements**	**SunScan mean measurements**
Seed	0	0	Seed	0	0
Germ	0	0	Germ	0	0
Appearance	0	0	Appearance	0	0
Closed lines	3	3	Closed lines	1	1
Blossoming	5	5	Blossoming	2	2
Tube growth	5	5	Fruit growth	2	2
Foliage ageing	5	5	Fruit maturation	2	2
Tube maturation	5	5	Foliage ageing	2	2
Foliage drying	2	2	Foliage drying	1	1
Phenological stages of *Groundnuts*	**GER1500 measurements**	**SunScan measurements**	**Phenological stages of *Beans***	**GER1500 mean measurements**	**SunScan mean measurements**
Seed	0	0	Seed	0	0
Germ	0	0	Germ	0	0
Appearance	0	0	Appearance	0	0
Closed lines	3	3	Closed lines	2	2
Tube growth	5	5	Blossoming	4	4
Foliage ageing	5	5	Fruit growth	5	5
Tube maturation	4	4	Fruit maturation	5	5
Foliage drying	1	1	Foliage ageing	2	2
			Foliage drying	1	1

In each sub-table, the phenological stages of each crop can be seen in the first column. In the 'GER 1500' column the number of spectroradiometric measurements taken at each stage are presented. For example, for Potatoes the measurements begun at the stage of 'closed lines' and there were 3 measurements during that stage (each measurement in the table is the average measurement from 25 measurements well spread in the plot. In the third column labelled 'Sunscan measurements' the LAI measurements are presented for each phenologi- cal stage as for the second column which were taken simultaneously. The number of each spectroradiometric measurement is not random. There should be a change in the reflectance in the specific phenological stage to have another measurement meaning that the crop reflectance during two consecutive days could be the same so the measurement would not enter the table as different one.

Changes in the phenological cycles of crops may occur from different parameters, such as weather conditions, soil and crop characteristics, and changes in the climate of an area (Minaccapili et al., 2008; Kross et al.,2011). Between years, phenological markers (such as length of growing season) may respond differently, a phenomenon which can be associat- ed with short-term climate fluctuations or anthropogenic forcing, such as groundwater extraction, urbanization (Bradley et al., 2007). However, the interpretation of phenological changes based on a large dataset volume for a period of many years can turn to be very complicated.

6. Semi-empirical modelling of satellite data (vegetation indices) to ground data (crop canopy parameters)

The commonly accepted equation for estimating evapotranspiration, according to the sche- matization of Monteith (Monteith and Unsworth, 1990), is a function of climate data such as temperature (T), humidity (RH%), solar radiation (R_s) and wind speed (U) and crop parame- ters, such as the surface albedo (*a*), the leaf area index (LAI) and the crop height (CH):

$$ETc = f\left(a, LAI, ch, T, RH\%, Rs, U\right) \tag{1}$$

Remote sensing techniques can be used for monitoring these vegetation characteristics. An analytical elaboration performed on Landsat reflectance values evidenced the possibility of retrieving the surface albedo (Brest and Goward, 1987), the leaf area index (Price, 1992) and the crop height (Moran and Jackson, 1991). Since these parameters directly affect the reflectance of cropped areas, it has been demonstrated that it is possible to establish a correlation between multispectral measurements of canopy reflectance and the corresponding canopy parameter's values (Bausch and Neale, 1987). In this study, the required crop parameters, *a*, LAI, CH have been derived from direct measurements and were correlated to reflectance measurements of the crops in each case.

Many studies have illustrated the need and the know-how for modeling or correlating LAI and Crop Height to remote sensing data and mainly to the vegetation indices inferred from handheld sensors. Leaf Area Index is an important structural property of crop canopy. High correlations were found between reflectance factor and LAI by Ahlrichs et al., (1983). Strong correlations between spectral data from crops and various characteristics of crops have been elucidated in numerous studies (Serrano et al, 2000;

Goel et al., 2003; Lee et al., 2004). Dar-vishzadeh et al., (2008) examined the utility of hyper spectral remote sensing in predicting canopy characteristics by using a spectral radiometer. Among the various models investigated, they found that canopy chlorophyll content was estimated with the highest accuracy. Some studies used multispectral image sensor system to measure crop canopy characteris- tics (Inoue et al., 2000)

Quantification of the canopy leaf area index (LAI) and its spatial distribution provides (Figure 7) an avenue to improve the interpretation of remotely sensed data over vegetat- ed areas. The purpose is to test the existing relation between vegetation indices with LAI and crop height and their prediction from remotely sensed data. It allow us to compare, on a consistent basis, the performance of a set of indices found in international litera- ture, in the prediction of LAI and CH which are basic parameters in the algorithms of estimating ETc. The method for mapping LAI and Crop Height for specific crops is shown in Figure 8.

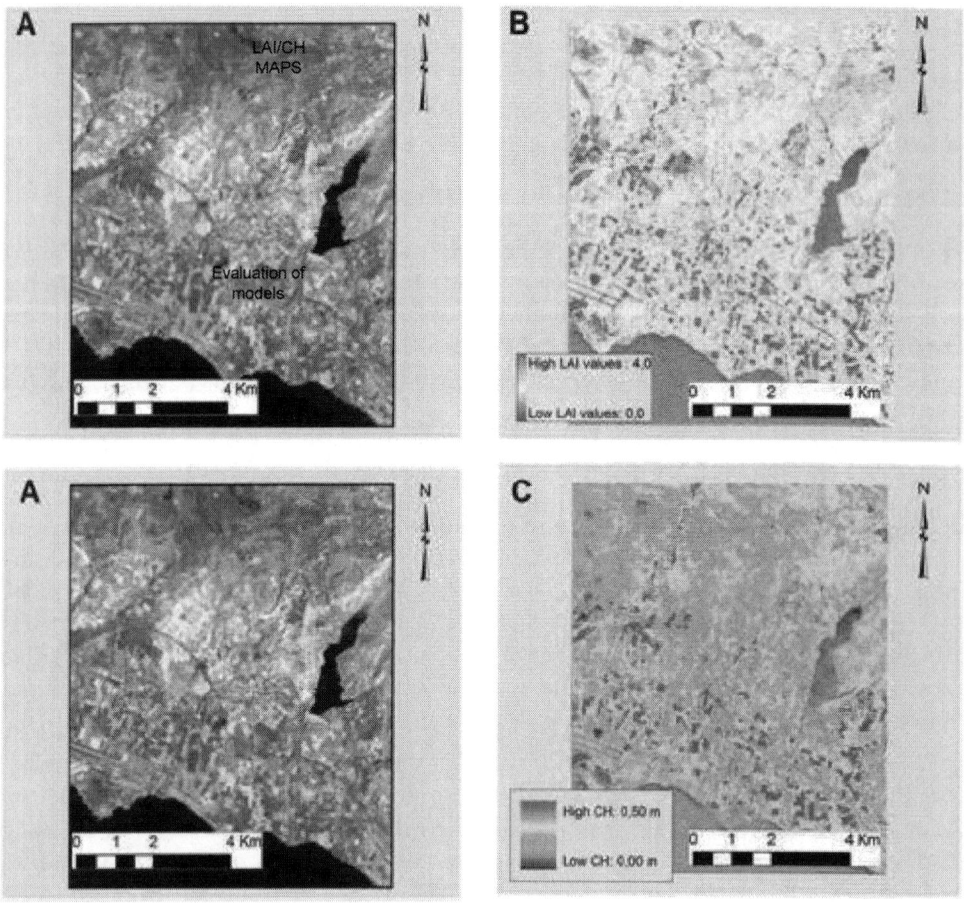

Figure 7. Production of LAI (B) and CH (C) maps (in pseudo color) using a Landsat image (A) (Papadavid, 2011)

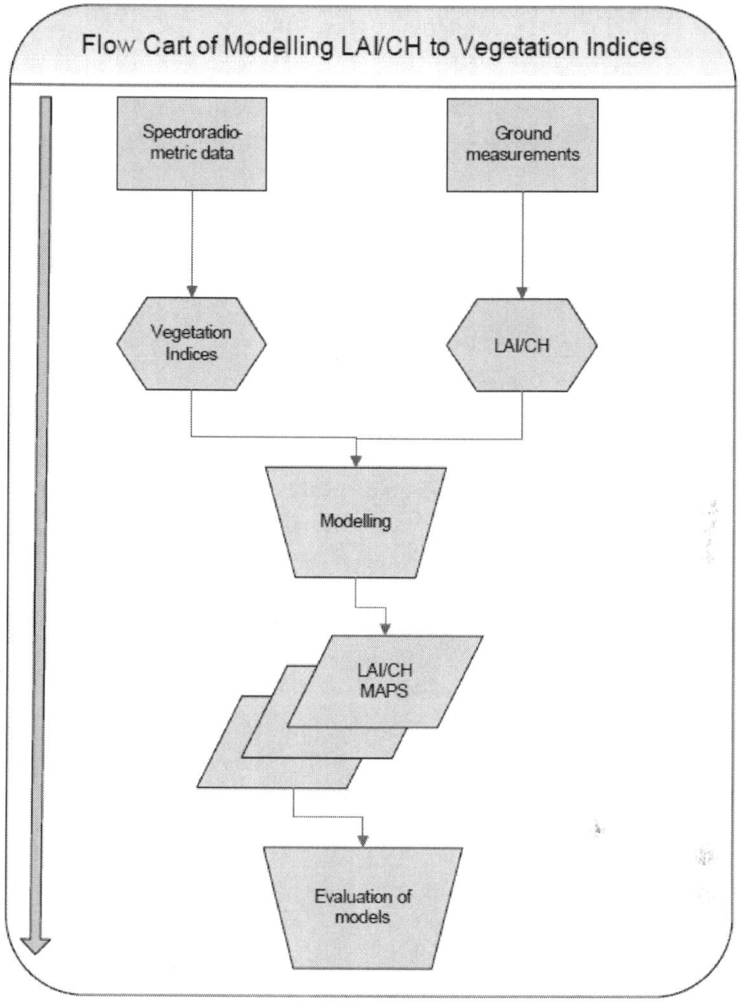

Figure 8. Method for mapping LAI and Crop Height using in situ and satellite remotely data

7. Algorithms application and results

7.1 SEBAL algorithm

SEBAL is a thermodynamically based model, using the partitioning of sensible heat flux and latent heat of vaporization flux as described by Bastiaanssen et al., (1998) who developed the algorithm. In the SEBAL model, ETc is computed from satellite images and weather data us- ing the surface energy balance as illustrated in Figure 9. Remotely sensed data in the visible, near-infrared and thermal infrared bands are used to derive the energy balance components along with ground measured solar radiation, if available. The other ground measurements that are required as model input are air temperature, relative humidity and wind speed at a point within the image.

SEBAL has an internal calibration for removing atmospheric effects using a series of iteration on Sensible Heat Flux (H) (Baastianssen et al., 2000; 2008). Since the satellite image pro- vides information for the overpass time only, SEBAL computes an instantaneous ET flux for the image time. The ET flux is calculated for each pixel of the image as a "residual" of the surface energy budget equation:

$$ET = R_n \quad G \quad H \tag{2}$$

where:

❖ R_n is the instantaneous net radiation ($W.m^{-2}$)

❖ G is the instantaneous soil heat flux ($W.m^{-2}$),

❖ H is the instantaneous sensible heat flux ($W.m^{-2}$)

❖ λET is the instantaneous latent heat flux ($W.m^{-2}$)

In this equation, the soil heat flux *(G)* and sensible heat flux *(H)* are subtracted from the net radiation flux at the surface *(R_n)* to compute the "residual" energy available for evapotranspiration *(λET)* (Figure 8). Soil heat flux is empirically calculated using vegetation indices, surface temperature and surface albedo. Sensible heat flux is computed using wind speed observations, estimated surface roughness and surface to air temperature differences. SEBAL uses an iterative process to correct for atmospheric instability due to the buoyancy effects of surface heating. Once the latent heat flux *(λET)* is computed for each pixel, an equivalent amount of instantaneous ET (mm/hr) is readily calculated by dividing by the la- tent heat of vaporization *(λ)*. Then, daily ETc is inferred.

When all parameters in Equation (2) are known, an instantaneous estimation of ETc can be conducted. Latent heat flux *(λET)* in Equation (3) is the rate of latent heat loss from the surface due to evapotranspiration, at the time of the satellite overpass. An instantaneous value of ETc_{inst} in equivalent evaporation depth is computed as:

$$ETc_{inst.} = 3600 \frac{ET}{} \tag{3}$$

where:

❖ ETc_{inst} is the instantaneous evapotranspiration (mm/hr)

❖ 3600 is the time conversion from seconds to hours

❖ λ is the latent heat of vaporization or the heat absorbed when a kilogram of water evaporates (J/kg)

The Reference ET Fraction (ET_rF) (Equation 4) is defined as the ratio of the computed instantaneous ET (ET_{inst}) for each pixel to the reference ET (ET_r) computed only from weather data:

$$ET_rF = \frac{ET_{inst}}{ET_r} \tag{4}$$

where:

❖ ET_r is the reference ET at the time of the image from the REF-ET software (mm/hr). ETrF is also known as crop coefficient, Kc. ETrF is used to extrapolate ET from the image time to 24-hour or longer periods. ET_rF values usually range from 0 to 1.

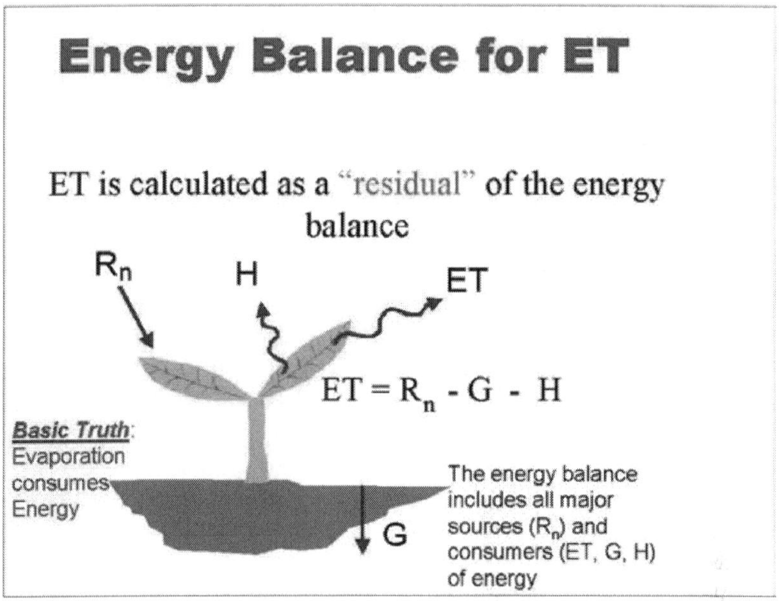

Figure 9. Energy Balance equilibrium (Source: Waters et al., 2002)

Finally, to get the daily values of ETc which are more useful than the instantaneous ones, SEBAL computes the ET_{daily} by assuming that the instantaneous ET_rF is the same as the 24- hour average. The daily ET_c (mm/day) is computed from Equation 5:

$$ET_c = ET_rF \times ET_r(24h) \quad (5)$$

where:

❖ ET_r (24h) is the total reference evapotranspiration of the day in mm/day.

Figure 10. ETc map of the area of interest (Landsat 5 TM image 2/1/2009) using SEBAL (Papadavid, 2011)

Daily ET_c is the final 'product' of SEBAL algorithm, meaning that satellite images are trans- formed into ET_c maps where one could retrieve ET_c for each pixel, as it is shown in Figure 10.

7.2 *Penman-Monteith* adapted to satellite data algorithm

Penman-Monteith method adapted to satellite data was used to estimate ET_c in mm/day (Equation 6). The specific equation needs both meteorological and remotely sensed data to be applied. The equation is used to estimate ET_c under some assumptions and depends on the direct application of the Penman-Monteith equation (Monteith, 1965; Rijtema, 1965; Smith, 1992; Allen et al., 1998) also based on EB theory, with canopy parameters estimated from satellite imagery (D'Urso et al., 2006; Minaccapili et al., 2008; Papadavid et al., 2010; 2011). Air temperature, atmospheric pressure, wind speed and other necessary meteorologi- cal data were collected from a weather station, located at the Paphos International Airport, very close to our study area. The method also needs empirical equations for describing the crop canopy factors (similar to SEBAL), namely albedo, crop height and LAI. It is a method with strong likelihood of correctly predicting the crop evapotranspiration in a wide range of tion has a strong theoretical basis, combining an energy balance to account for radiation and sensible heat transfer with an aerodynamic transport function to account for transfer of va- por away from the evaporating surface. The method is described as follows:

$$ETc = \frac{(R_n \quad G) + c_p p_a \left(e_s \quad e_a\right) / r_{ah}}{+ \left(1 + r_s / r_{ah}\right)} \tag{6}$$

or

$$ETc = \frac{86400}{\lambda}\left[\frac{\Delta(R_n - G) + c_p p_a\left(e_s - e_a\right) / r_{ah}}{\Delta + \gamma\left(1 + r_s / r_{ah}\right)}\right]$$

where

- ❖ ETc is the crop evapotranspiration (mm/day)
- ❖ Δ represents the slope of the saturated vapor pressure temperature relationship (kPa / K1)
- ❖ R_n is the net solar radiation (W/m^2)
- ❖ G Soil Heat flux (W/m^2)
- ❖ c_p is the air specific heat (J/kg K)
- ❖ ρa is the air density (kg / m^3)
- ❖ e_s is the saturated vapor pressure (kPa)
- ❖ e_a is the actual vapour pressure (kPa)
- ❖ r_{ah} is the aerodynamic resistance (s/m)
- ❖ r_s is the surface resistance (s/m)

- ❖ λ is the latent heat of vaporisation of water (J / kg)
- ❖ γ is the thermodynamic psychrometric constant (kPa / K)

This equation is valid under conditions of intense solar irradiance (typical summer condition in Mediterranean climate) and for 0,5 < *LAI* < 3, which is the case for Cyprus annual crops. What is important in the specific model is that of its use without the need of the ther- mal band of any satellite, contrary to the other Energy Balanced based models which ther- mal band is a prerequisite (Papadavid et al., 2011). Another difference that is rising in this model compared to SEBAL, is the need of atmospheric corrections where SEBAL and other models have an internal calibration for compensating atmospheric effects. The parameters Δ, G, u_2, e_s-e_a, R_n and Δ are calculated according to the formulae of the method by the con- ventional data of the meteorological station situated in the area of interested. The formulae for calculating each parameter can be found extensively in 'FAO Irrigation and Drainage pa- per No. 56' by FAO (1998). As in all ETc algorithms the final product is an ETc maps (Figure 11) of the area of interest where users can infer the ETc values for specific crops.

Figure 11. ETc map of the area of interest (Landsat 7 TM image 2/1/2009) using P-M (Papadavid, 2011)

The results regarding crop water requirements of the different crops can be found on Table 3. The water needs for each crop is average value, for each month, based on the crop evapo- transpiration found employing the two algorithms described above, after applying the methodology for modeling crop parameters to satellite data.

Table 3. Crop Water Requirements for the different crops (m3/ha/month)

Crop	J	F	M	A	M	J	J	A	S	O	N	D	Total
Potatoes							450	850	1200	1550	1300		5330
Ground nuts							620	1450	1650	300			4020
Beans							450	850	1200	990			3490
Chick peas			200	800	480								1480

8. Use of wireless sensors for supporting evapotranspiration measurements and smart management of irrigation demand

Wireless sensors have been used in this study as an extra tool for supporting evapotranspiration measurements in the same area of interest (Hadjimitsis et al., 2008a & 2008b). Such sensors were used as smart meteorological stations (relative humidity, temperature, wind speed) as well as tools for retrieving soil moisture, soil temperature leaves wetness and temperature. These information can be used to assess our evapotranspiration results.

Figure 12. Wireless nodes in the Mandria area in Paphos (Papadavid, 2012)

The Wireless Sensor Network (WSN) consisted of a number of wireless nodes (near to 20 nodes) placed at various locations in the surrounding agriculture fields irrigated from the Asprokremmos Dam in Paphos District area in Cyprus (see Figure 12). The WSN acts as a wide area distributed data collection system deployed to collect and reliably transmit soil and air environmental data to a remote base-station hosted at Cyprus University of Technology (at the Remote Sensing Laboratory), as shown in Figure 13.

The micro-sensors were deployed using ad-hoc multi-hop communication protocol and transmit their data to a gateway which is responsible to collect, save and forward them to a remote database through a GPRS connection. The solar powered gateway, shown in Figure 14, was equipped with various meteorology sensors required to assist the indeed research project such as rain, wind, barometric pressure, temperature etc, which give ad- ditional information to the system. The gateway also hosts a GPS sensor for identifying the exact position of the WSN an event-driven smart camera for acquiring real-time pic- tures of the area and also a GPRS modem for communicating with the remote server which is deployed tens of thousands of kilometers away. The absence of power and com- munication infrastructure was tackled by creating a fully solar operated gateway (autono- my of three days without sunshine) and by incorporating a low power GPRS modem for communication. A multi-parameter decision system running on the remote server would be able to process the sensor data and produce valuable information about watering dif- ferent vegetables and create early notifications and suggestions which are then distribut- ed to farmers and water management authorities. The system was able to process multi parameter data collected from different sensors such as soil moisture, soil temperature (Figures 15 and 16), leaves wetness and temperature, humidity, rainfall, wind speed and direction and ambient light.

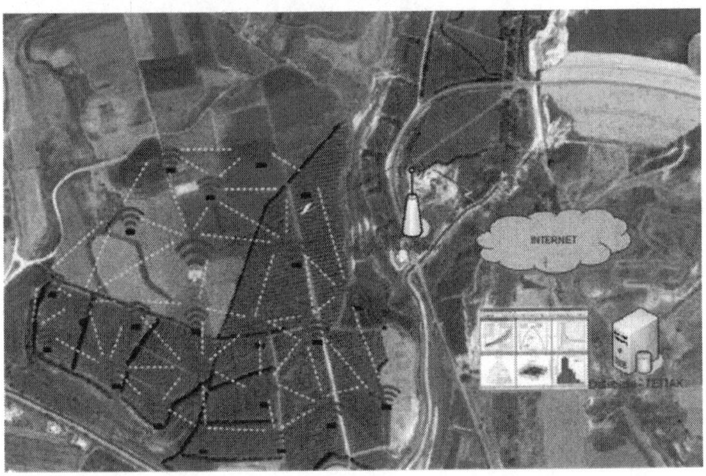

Figure 13. Wirelesses Sensor Network Schema (Hadjimitsis et al., 2008a & 2008b)

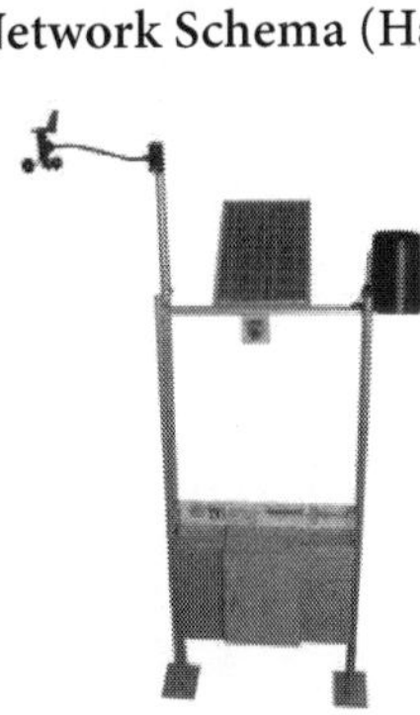

Figure 14. The WSN Gateway and Meteorology Station (Hadjimitsis et al., 2008a & 2008b)

Figure 15. Soil moisture measurements using micro-sensor technology in agricultural field

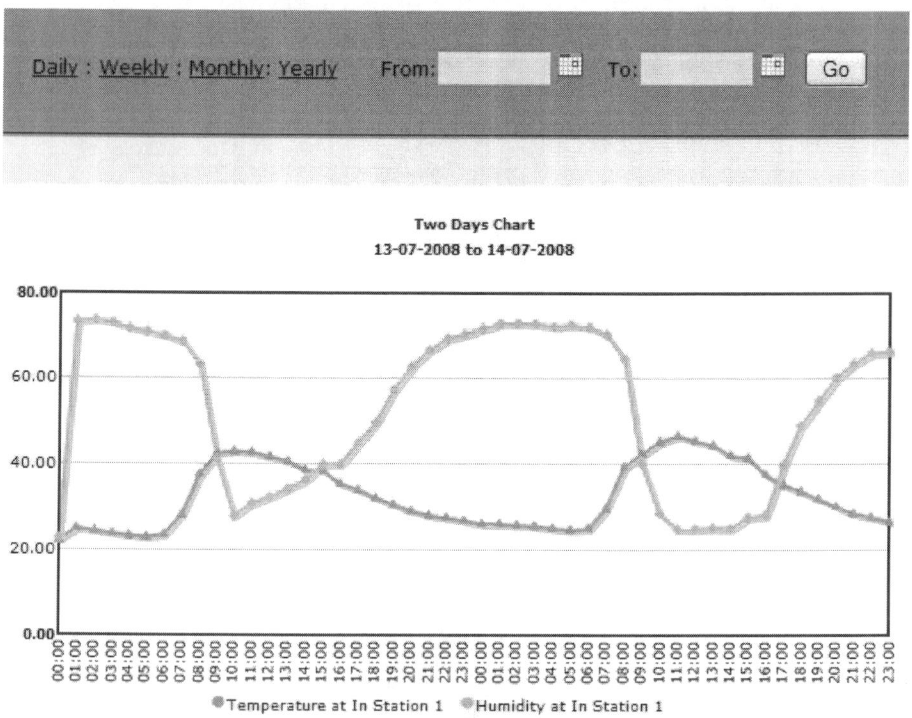

Figure 16. Temperature measurements coming from MSN.

9. Smart management of evapotranspiration using 3G telephony

As ETc is calculated each night based on that day's both weather readings and satellite images using the previously described method, the ETc results from these calculations are sent to farmers each morning giving them the water balance (Crop water requirements) for their area for the irrigation season until the previous day. Using the existing method by combining satellite-derived crop coefficients and the 3G telephony with SMS delivery service, now offers the potential to provide low cost, site specific and personalised (for crop type and management conditions) irrigation water management information to individual farmers across an irrigation region (Papadavid et al., 2012). Automatically triggered text messages can be generated by server-based software that combine data and formatting and then send the message out to mobile phones via an Internet cellular network gateway services. 3G phones can not only send SMS but can also send extended multimedia.

High spatial resolution of water management information (approx. 30 m x 30 m using Land- sat 5) allows farmers to better manage spatial variability to maximise production, minimise costs and environmental impacts.

10. Concluding remarks

As a key component in water resources management, it is essential to estimate evapotranspiration accurately for water resources evaluation, drought monitoring and crop production simulation. Accurate estimates of ETc are needed for numerous agricultural and natural resource management procedures. However, this is difficult to achieve in practice because actual evapotranspiration cannot be measured directly and varies considerably in time and space.

Satellite images are collected across Mediterranean areas with frequencies ranging from daily to monthly. The clear skies enable the gathering of good quality information and it is now possible to use satellite remote sensing to estimate the rates of ETc as shown in this chapter. Research has shown that there is a direct relationship between vegetation cover such as indices and ETc. This means that the standard approach of using static crop water requirement look-up tables can be improved by using the more dynamic and customised information provided by satellite imagery. Satellite Remote sensing can assist in improving the estimation of ETc, and consequently water demand in cultivated areas for irrigation purposes and sustainable water resources management.

In this Chapter remotely sensed data along with meteorological data, modeling techniques and surface energy balance algorithms were combined. All these procedures combined can provide the spatial distribution of ETc in maps where users can derive the value of ETc for each crop in mm/day. The methodology followed can be applied for any place since it can be considered as 'algorithm adaptation' to local conditions. The parameters that are required in the empirical equations can be easily evaluated using remote sensing techniques and field spectroscopy. Modeling techniques (for example, regression analysis) are used to correlate and evaluate measured crop canopy factors, such as Leaf Are Index (LAI) and Crop Height (CH), to remotely sensed data uring the entire phenological cycle of each crop. The intention is to create semi-empirical models describing LAI and CH, which are indispensible parameters in almost all ETc algorithms, using remotely sensed data. Using these models, users can avoid direct measurements of these parameters every time there is an application of an ETc algorithm.

The methodology as described in this chapter can support decision makers of Water Authorities. The methodology was applied for Landsats' images but it can easily be adapted for other satellite sensors. The use of field-spectroradiometer can facilitate the procedure since it provides a spectrum which can be adapted to satellites' bands by simple transformation, using relative spectral response (RSR) filters of each satellite.

References

1. Ahlrichs, J.S. and M.E. Bauer. 1983. Relation of agronomic and multispectral reflectance characteristics of spring wheat canopies. Agronomy Journal 75: 987– 993.
2. Alexandridis T. 2003. Effect of scale on hydrological and vegetation parameters estimation using remote sensing techniques and GIS, PhD study, Aristotle University of Thessaloniki, Greece.
3. Allen R.G., Pereira L.S., Raes D. and Smith M. 1998. Crop evapotranspiration, guidelines for computing crop water requirements. FAO Irrigation and Drainage Paper 56, Food and Agricultural Organization of United Nations (FAO) Rome, Italy, pp 300.
4. Ambast S.K., Ashok K, Keshari, Gosain A.K. 2006. Satellite remote sensing to support management of irrigation systems: concepts and approaches. Irrigation and Drainage systems 51:p 25-39.
5. Anderson K., Milton E.J. and Rollin E.M. 2006. Calibration of dual-beam spectroradiometric data. International Journal of Remote Sensing, 27, 975–986.
6. Bastiaanssen W.G.M. 1995. Regionalization of surface flux densities and moisture indicators in composite terrain, Doctoral thesis, Agricultural University, Wageningen, The Netherlands, pp 273.
7. Bastiaanssen W.G.M. 2000. SEBAL-based sensible and latent heat fluxes in the irrigated Gediz Basin, Turkey. Journal of Hydrology 229:87-100.
8. Bastiaanssen W.G.M. and Ali S. 2003. A new crop yield forecasting model based on satellite measurements applied across the Indus Basin, Pakistan. Agriculture, Ecosystems and Environment, 94:321-340.
9. Bastiaanssen W.G.M., Brito Bos M.G., Souza K.A., Cavalcanti E.B. and Bakker M.M. 1998. Low cost satellite data for monthly irrigation performance monitoring: Irrigation and Drainage systems 15 : p 53-79.
10. Bastiaanssen W.G.M., Menenti M., Feddes R.A. and Holtslag A.A.M. 1998. A remote sensing surface energy balance algorithm for land (SEBAL), part 1: formulation, Journal of Hydrology. 212-213: 198-212.
11. Bastiaanssen W.G.M., Noordman E.J.M., Pelgrum H., David G., Thoreson B.P. and Allen R.G. 2005. SEBAL model with remotely sensed data to improve water-resour- ces management under actual field conditions. ASCE J. Irrig. Drain. Eng. 131(1): 85-93.
12. Bausch, W.C. and Neale, C.M.U. 1987. Crop coefficients derived from reflected canopy radiation: a concept. *Transactions American Soc. Agric. Engin.* 30 (3): 703-709.

13. Beisl, U. (2001). Correction of bidirectional effects in imaging spectrometer data. Re- mote Sensing Series, Vol. 37. Zürich, Switzerland Remote Sensing Laboratories, Uni- versity of Zürich.

14. Bradley A.B, Jacob W.R., Hermance F.J., Mustard F.J. 2007. A curve fitting procedure to derive inter-annual phenologies from time series of noisy satellite NDVI data, Remote Sensing of Environment, 106 (2), 137-145.

15. Brest, C.L. and Goward, S.N. 1987. Deriving surface albedo measurements from narrow band satellite data. *Int. J. Remote Sensing* 8 (3): 351-367.

16. Bruegge C.J., Chrien N. and Haner D. 2001. A Spectralon BRF database for MISR cali- bration applications. Remote Sensing of Environment, 76, 354–366.

17. Courault D., Seguin B. and Olioso A. 2005. Review on estimation of Evapotranspiration from remote sensing data: from empirical to modeling approaches. Irrigation and Drainage systems 19: p. 223-249.

18. D'Urso G., Calera Belmonte A. 2006. Operative approaches to determine crop water requirements from Earth Observation data: methodologies and applications. In D'Ur- so G., Osann Jochum M.A., Moreno J. (Eds.): Earth Observation for vegetation moni- toring and water management", Conference Proceedings Naples. 2005. 9-10 Nov., published by American Inst. Physics, Vol. 852: 14-25.

19. D'Urso G., Querner E.P. and Morabito J.A. 1992. Integration of hydrological simulation models with remotely sensed data: an application to irrigation management. Leuven, Belgium , p:463-472.

20. Dadhwal V.K., Parihar J.S. and Medhavy T.T. 1996. "Comparative performance of thematic mapper middle-infrared bands in crop discrimination" Int. J. of Remote Sensing, 17(9), pp. 1727-1734.

21. Dangel S., Kneubühler M., Kohler R., Schaepman M., Schopfer J., Schaepman-Strub G., et al. 2003. Combined Field and Laboratory Goniometer System — FIGOS and LAGOS. International Geoscience and Remote Sensing Symposium (IGARSS), 7, 4428–4430.

22. Darvishzadeh, R., A. Skidmore, M. Schlerf, C. Atzberger, F. Corsi and M. Cho. 2008. LAI and chlorophyll estimation for a heterogeneous grassland using hyperspectral measurements. *ISPRS Journal of* Photogrammetry & Remote Sensing 63(4): 409- 426.

23. Funk C., Budde E.M. 2009. Phenologically-tuned MODIS NDVI-based production anomaly estimates for Zimbabwe, Remote Sensing of Environment, 113 (1), 115-125.

24. Goel, P.K., S.O. Prasher, J.A. Landry, R.M. Patel and A.A. Viau. 2003. Estimation of crop biophysical parameters through airborne and field hyperspectral remote sensing. Transactions of the ASAE 46(4): 1235–1246.

25. Gouranga K., Harsh N.V. 2005. Phenology based irrigation scheduling and determination of crop coefficient of winter maize in rice fallow of eastern India, Agricultural Water Management, 75 (3), 169-183.

26. Hadjimitsis, D.G., Clayton C.R.I., Retalis A. and Spanos K. 2000. Investigating the po- tential of using satellite remote sensing for the assessment of water quality in large

dams, and irrigation demand, in Cyprus. Proceedings 26th Annual Conference and Exhibition of the Remote Sensing Society, RSS2000 , University of Leicester.

27. Papadavid G., Perdikou S., Hadjimitsis M.G., Hadjimitsis, D.G. 2012. Smart management and monitoring of irrigation demand in Cyprus using remote sensing and 3rd generation mobile phones, 32nd EARSeL Symposium 2012, Mykonos, Greece.

28. Hadjimitsis D.G., Hadjimitsis M.G, Toulios L. and. Clayton C.R.I. 2010. Use of space technology for assisting water quality assessment and monitoring of inland water bodies, Journal of Physics and Chemistry of the Earth, 35 (1-2), pp. 115-120, DOI: 10.1016/j.pce.2010.03.033

29. Hadjimitsis D.G.; Papadavid G.; Agapiou A.; Themistocleous K.; Hadjimitsis M.G.; Retalis A., Michaelides S.; Chrysoulakis N.; Toulios L. and Clayton C.R.I. (2010). At- mospheric correction for satellite remotely sensed data intended for agricultural ap- plications: impact on vegetation indices, *Nat. Hazards Earth Syst. Sci.*, 10, 89-95, doi: 10.5194/nhess-10-89-2010.

30. Hadjimitsis D.G., Kounoudes A. and Papapadavid G. 2008a. Integrated Method for Monitoring Irrigation Demand in Agricultural fields in Cyprus using satellite remote sensing and wireless sensor network. 4th International Conference HAICTA 2008 Proceedings -'Information and Communication Technologies in Bio & Earth Sciences',18-20/9/2008, Agricultural University of Athens, Editor: T. Tsiligiridis, ISBN 978-960-387-725-7, p.10-16.

31. Hadjimitsis D.G., Papadavid G., Themistocleous K., Kounoudes A., and Toulios L. Estimating irrigation demand using satellite remote sensing: a case study of Paphos District area in Cyprus. 2008b. Remote Sensing for Agriculture, Ecosystems, and Hy- drology X. Edited by Neale, Christopher M. U.; Owe, Manfred; D'Urso, Guido. Pro- ceedings of the SPIE, Volume 7104, pp. 71040I-71040I-11, Proceedings of SPIE Europe Remote Sensing, 15 - 18 September 2008 University of Wales Institute, Cardiff, UK. DOI: 10.1117/12.800366

32. Hoedjes J.C.B., Chehbouni A., Jacob F., Ezzahar J. and Boulet G. 2008. Deriving daily Evapotranspiration from remotely sensed evaporative fraction over olive orchard in Morocco. Journal of Hydrology: 53-64.

33. Inoue, Y., S. Morinaga and A. Tomita. 2000. A blimp–based remote sensing system for low–altitude monitoring of plant variables: A preliminary experiment for agricul- ture and ecological applications. *International Journal of Remote Sensing* 21(2): 379– 385.

34. Jackson R.D., Clarke T.R. and Moran M.S. 1992. Bi-directional calibration results for 11 Spectralon and 16 BASO4 reference reflectance panels, Remote Sensing of Environment, 40, 231-239.

35. Kross A., Fernandes R., Seaquist J., Beaubien E. 2011. The effect of the temporal resolution of NDVI data on season onset dates and trends across Canadian broadleaf for- ests, Remote Sensing of Environment, 115 (6), 1564-1575.

36. Lang A., McMutrie R., Benson M. 1991. Validity of surface area indices of Pinus Radiata estimated from transmittance of the sun's beam. Agric. Forest Meteo. 57: 157-170.

37. Lee, K.S., W.B. Cohen, R.E. Kennedy, T.K. Maiersperger and S.T. Gower. 2004. Hyperspectral versus multispectral data for estimating leaf area index in four different biomes. *Remote Sensing Environment* 91(3-4): 508–520.

38. Maier S.W. 2000. Modeling the radiative transfer in leaves in the 300 nm to 2.5 μm wavelength region taking into consideration chlorophyll fluorescence - The leaf model SLOPE, PhD Thesis, Deutsches Fernerkundungstagsdatenzentrum, Technische Universidad München, Oberpfaffenhofen (Germany), 110 pp.

39. McCloy K.R. 2010. Development and Evaluation of Phenological Change Indices De- rived from Time Series of Image Data, Remote Sensing, 2, 2442-2473.

40. McLoy K.R. 1995. Resource Management information systems, Taylor and Francis, London, 244-281.

41. Minacapilli M., Iovino M., D'Urso G. 2008. A distributed agro-hydrological model for irrigation water demand assessment. Agricul t u r a l water management 95, 123 – 132.

42. Monteith J.L. 1965. Evaporation and the environment. In: The state and movement of water in living organisms. 19th Symp. Soc. Biol., pp. 205 – 234.

43. Monteith J.L. and Unsworth M.H. 1990. Principles of Environmental Physics, Second Edition, Butterworth Heinemann. ISBN 0-7131-2931- X.

44. Moran M.S., Inoue Y. and Barners E.M. 1997. Opportunities and limitations for image-based remote sensing in Precision Crop Management: in RS Environment 61, p 319-349.

45. Newnham G.J. and Burt T. 2001. Validation of a leaf reflectance and transmittance model for three agricultural crop species, in Proc. International Geoscience and Remote Sensing Symposium (IGARSS'01), Sydney (Australia), IEEE, Vol. 7, pp. 2976 -2978.

46. O'Leary G.J., Connort D.J., White D.H. 1985. A Simulation Model of the Development, Growth and Yield of the Wheat Crop, Agricultural Systems 17, 1-26.

47. Papadavid G., Hadjimitsis D. (2010). An integrated approach of Remote Sensing techniques and micro-sensor technology for estimating Evapotranspiration in Cyprus. *Agricultural Engineering International: CIGR Journal*, Manuscript 1528, Vol. 12, No. 3.

48. Papadavid G.; Agapiou A.; Michaelides S. and Hadjimitsis D.G. (2009). The integration of remote sensing and meteorological data for monitoring irrigation demand in Cyprus. *Nat. Hazards earth syst. Sciences*, 9, 2009-2014.

49. Papadavid G.; Hadjimitsis D.; Toulios L., Michaelides L. (2011). Mapping Potatoes Crop Height and LAI through Vegetation Indices using Remote Sensing, in Cyprus *Journal of Applied Remote Sensing* 5, 053526 (2011), DOI:10.1117/1.3596388.

50. Papadavid G.; Hadjimitsis D.G., Michaelides S. (2011). Effective irrigation management using the existing network of meteorological stations in Cyprus. *Advances in Geosciences Journal*, 9, 7-16, doi:10.5194/adgeo-30-31-2011.

51. Papadavid G.; Hadjimitsis D.G.; Kurt Fedra and Michaelides S. (2011). Smart management and irrigation demand monitoring in Cyprus, using remote sensing and water resources simulation and optimization. *Advances in Geosciences Journal*, 9, 1-7, doi:10.5194/adgeo-9-1-2011.

52. Papadavid G.; Hadjimitsis D.G.; Perdikou S.; Michaelides S.; Toulios L.; Seraphides N. (2011). Use of field spectroscopy for exploring the impact of atmospheric effects on Landsat 5 TM / 7 ETM+ satellite images intended for hydrological purposes in Cyprus, *GIScience and Remote Sensing*, 48, No 2, p. 280—298, DOI: 10.2747/1548-1603.48.2.280.

53. Papadavid G.;. (2012). Estimating evapotranspiration for annual crops in Cyprus us- ing remote sensing. *Phd Thesis, Department of Civil Engineering and Geomatics, Cyprus University of Technology, Lemesos, Cyprus.*

54. Price, J.C. 1992. *Estimating Leaf Area Index from Remotely Sensed Data.* Proc. IGARSS '92 (Houston). Vol. 1. pp. 1500-1502.

55. Rijtema P.E. 1965. An analysis of actual evapotranspiration. Agric. Res. Rep., 659, Pu- doc, Wageningen, pp107.

56. Rogers D. and Alan M. 2007. An Evapotranspiration Primer. Irrigation Management Series. Kansas.

57. Schaepman M.E. 2007. Spectrodirectional remote sensing: From pixels to processes. International Journal of Applied Earth Observation and Geoinformation, 9(2), 204–223

58. Serrano, L., I. Filella and J. Penuelas. 2000. Remote sensing of biomass and yield of winter wheat under different nitrogen supplies. Crop Science 40: 723–731.

59. Shaykewich C.F. 1994. An appraisal of cereal crop phenology modelling, Canadian Journal of Plant Science, 329-341.

60. Smith M. 1992. CROPWAT, a computer program for irrigation planning and management. Irrigation and Drainage Paper 46, FAO, Rome, Italy.

61. Song, J.; Duanjun, L.; Wesely, M.L. 2003. A simplified Atmospheric Correction Procedure for the Normalized Difference Vegetation Index. *Photogrammetric Engineering & Remote Sensing*, 69, 521–528.

62. Telis A. and Koutsogiannis D. 2007. Estimation of Evapotranspiration in Greece. PhD Thesis, Athens.

63. Thiruvengadachari S. and Sakthivadivel K. 1997. Satellite remote sensing for assessment of irrigation system performance. Research Report 9, IWMI Colombo, Srilanka.

64. Tsouni A. and Koutsogiannis D. 2003. The contribution of remote sensing techniques to the estimation of Evapotranspiration : the case of Greece. PhD Thesis, Athens.

65. Wang L., Wang W., Dorsey J., Yang X., Guo B. and Shum H.Y. 2005. Real-time rendering of plant leaves, in Proc. ACM SIGGRAPH 2005, Los Angeles (USA), 31 July - 4 August 2005, pp. 167-174.

66. Welles and Norman. 1991. Instrument for measurement of canopy architecture. Agron J. 83: 818-825.

Chapter-3

Remote Sensing for Archaeological Applications: Management, Documentation and Monitoring

1. Introduction

Archaeology is defined as the systematic approach for uncovering the human past and its environment. Archaeology involves not only systematic excavations and surveys, but also analysis of the data collected in the field. In a broader term, archaeology is an interdisciplinary research. Modern studies in archaeology engage a series of other sciences such as geology, information systems, chemistry, statistics, etc. In recent years, remote sensing has received considerable attention since it can assist archaeological research, along with other sciences, in order to extract valuable information to the researchers based only on non-destructive and non-contact techniques.

Remote sensing is the acquisition of information about an object or phenomenon without making any physical contact with the object (Levin, 1999; Parcak, 2009). According to Sabins (1997), remote sensing involves all the methods that allow the use of electromagnetic radiation in order to identify and detect various phenomena. Based on this definition, many techniques such as satellite remote sensing, aerial photography, geophysical surveys, ground spectroscopy or even terrestrial laser scanners, are considered as remote sensing techniques (Johnson, 2006).

Remote sensing has opened up new horizons and possibilities for archaeology. For example, oblique or vertical aerial photography can detect phenomena on the surface associated with subsurface relics, while the use of infrared and thermal electromagnetic radiation can be used in order to detect underground archaeological remains (Bewley et al., 1999; McCauley et al., 1982). Moreover, remote sensing as a non-destructive technique can contribute to the investigation of an archaeological site before, during and after excavation periods. At the micro-level scale, geophysical surveys and ground spectroscopy can provide information about subsurface relics, while at the macro-scale, aerial photographs and satellite remote sensing can identify traces of the human past. Concurrently, these techniques can monitor the surroundings of a cultural heritage site and record any changes due urban expansion and/or changes of land use (Rowlands & Sarris, 2007; Ma- sini & Lasaponara, 2007; Hadjimitsis et al., 2009;

Ventera et al., 2006; Negria & Leucci, 2006; Cavalli et al., 2007; Altaweel 2005; Aqdus et al., 2008; Bassani et al., 2009).

Satellite remote sensing has become a common tool of investigation, prediction and forecast of environmental change and scenarios through the development of GIS-based models and decision-support instruments that have further enhanced and considerably supported deci- sion-making (Ayad, 2005; Douglas, 2005; Hadjimitsis et al., 2006; Cavalli et al., 2007). By blending together satellite remote sensing techniques with GIS, the monitoring process of archaeological sites can be efficiently supported in a reliable, repetitive, non-invasive, rapid and cost-effective way (Hadjimitsis and Themistocleous, 2008).

This chapter presents a brief overview of the evolution of remote sensing in archaeological research. Several applications of applied remote sensing techniques, including satellite remote sensing, GIS, laser scanning, atmospheric pollution, spectroscopy, webGIS and ge- ophysical prospection will also be examined through different case studies in Cyprus and Greece.

2. Satellite remote sensing in archaeology

This section introduces current remote sensing satellite data which are available for archaeological research along with a historical background of remote sensing applications in archaeology. As well, satellite sensors, such as Landsat, EO – Hyperion, QuickBird, IKONOS, etc., are also briefly outlined.

2.1 Historical review

The first aerial photographs used for archaeological purposes were taken just before the be- ginning of World War I in UK and Italy (Capper, 1907; Parcak, 2009; Bewley et al., 1999; Riley, 1987). Mesopotamia and the Levant were traditionally photographed until the 1940s (see Keneddy, 1925; Crawford, 1923, Glueck, 1965, Keneddy, 2002). After the end of World War II, new archaeological sites were explored due to aerial reconnaissance during the war. The scientific interest has been currently shifted to the Middle and Far East, as well as other areas in Europe and America (Parcak, 2009). During the Cold War in the 1960's, several sat- ellites, including CORONA, Argo, Lanyard and COSMOS, were used for military purposes. However, these data were only accessible after their declassification in 1995 (Parcak, 2009).

Spatial resolution of CORONA spy images taken during the Cold War could reach up to 0.6m (Lock, 2003). Fowler & Fowler (2005) explored the potentials of CORONA images for archaeological purposes and concluded that such images can be used as an alternative way in many European archaeological sites, where traditional aerial photography is very limited. Grosse et al., (2005) used CORONA images for mapping geomorphological features in NE Siberia. The combination of ASTER and CORONA images in northern Mesopotamia was also studied by Altaweel (2005).

KVR-100 images from the Russian space program have been available since 1987 and have a high spatial resolution of 2-3 m. Such data are valuable in areas where the landscape has changed dramatically as a result of human activity, such as urban expansion. Even though KVR-100 has been used by several researchers (Fowler and Curtis, 1995; Comfort, 1997), their application is still limited due to their high cost (Parcak, 2009). CORONA and KVR images have been also used to monitor cultural heritage sites in Iran (Kostka, 2002).

Since the 1970s, the launch of new satellite systems coincided with the technological progress of the sensors. In 1972, the Landsat space program was initiated and was followed by the launch of other satellites, including the SPOT satellite in France (Parcak, 2009; Sarris, 2008). The Landsat sensor has been in continuous orbit since 1972 and provides multispectral data for archaeological research. Despite the medium spatial resolution (from 15-80m) Landsat images have a relatively low cost while covering a large area (180 x 180 km) in both the visible - infrared and thermal wavelengths. Landsat images were used to study archaeolandscapes in many archaeological projects and surveys. Vaughn and Crawford (2009) used predictive models in order to identify new areas with potential settlements of Mayans. Barlindhaug et al., (2007) found that Landsat satellite images can be used for monitoring purposes of archaeological sites. Neolithic settlements in Greece were detected using archive Landsat images (Alexakis, 2009; Agapiou et al., 2012a; 2012b). Landsat images were also used for monitoring purposes of the surroundings of monuments in Cyprus (Hadjimitsis et al., 2009; 2008).

During the 1980's, thermal and radar sensors were also added to satellite sensors (Bewley et al., 1999). In the late 1980's, India launched the IRS 1A, 1B, 1C, 1D and IRS P2 sensors (Tripa- thi 2005a). Although these data have been used for archaeological purposes in India, such as the identification of the mythic site *Dvaraka* (Tripathi 2005b) and the observation of *Hampi* site (Raj et al., 2005), their use is very limited in other regions.

From the 1990's, remote sensing and Geographical Information Systems (GIS) have been used systematically for archaeological research and newer satellites with higher spatial reso- lution are now available. Indeed, Quickbird, IKONOS, WorldView and GeoEye are capable of providing satellite images with spatial resolution up to 0.5 m.

In addition to the above, hyperspectral images, such as those from EO-HYPERION, have recently made their appearance. Hyperspectral remote sensing analysis is performed over hundreds of narrow bands. The key characteristics of hyperspectral images are its fine spectral and radiometric resolution. Hyperspectral data provides a variety of spectral information, which can be used for the identification of archaeological remains. Alexakis et al., (2009) stated that these new technologies can support the detection of archaeological sites, although it is not always possible to extract a unique archaeological spectral signature due to the heterogeneous presence of vegetation and soil.

Lasaponara and Masini (2007a) highlighted the potential benefits of high resolution satellite images in order to detect subsurface monuments through the use of vegetation indices and edge detection techniques. Cavalli et al., (2007) introduced the use of airborne hyperspectral scanner Multispectral Infrared Visible Imaging Spectrometer (MIVIS) for the detection of subsurface monuments based on spectral anomalies. The study found that the detection of subsurface monuments is possible employing both visible and near infrared part of electro- magnetic radiation, and can concurrently detect anomalies using the thermal infrared spectrum. Using QuickBird satellite imagery, Lasaponara and Masini (2007b) examined the Metaponto archaeological sites in the South of Italy, using sophisticated spectral techniques such as the Tasselled Cap Transformation and Principal Component Analysis. The combina- tion of hyperspectral data and several remote sensing processing techniques (Principal Component Analysis, vegetation indices, etc.) for the detection of subsurface monuments in eastern Scotland was also presented by Aqdus et al., (2009).

Beck (2007) and Beck et al., (2007) conducted a detail study of the archaeological site of *Homs* in Syria, using CORONA and IKONOS images. The results indicated that areas with archaeological interest tend to have different spectral signatures from the surrounding area. Rowlands and Sarris (2007) used airborne hyperspectral scanners (Airborne Thematic Mapper –ATM and Compact Airborne Spectrographic Imager -CASI) and LIDAR data in order to study the Hellenistic settlement of *Itanos* in Crete. The data were post-processed using object-oriented analysis. Although the study found several difficulties in relation to the identification of archaeological remains, the continuing use of such methods and applications along with other remote sensing techniques such as geophysical surveys was recommended. In the ancient city *Sagalassos*, Laet et al., (2007) applied object-oriented techniques and several satellite images (ASTER, SPOT, IKONOS) in order to identify archaeological remains. The results from investigations , in the Piramide Nar- anjada in Cahuachi (Peru), based on high resolution satellite imagery, geomagnetic sur- veys and Ground Probing Radar was recently presented by Lasaponara et al., (2011). Currently, several archaeological investigations are carried out using combined remote sensing techniques, such as satellite images, aerial photographs, ground geophysical surveys, and LIDAR measurements. The next section provides an outline of the characteristics of the most important satellite data available today for archaeological research.

2.2 Satellite image data

Currently, there is a plethora of satellite images which may be used for supporting archaeo- logical research. However, these images have different resolutions depending on the sensor characteristics. Moreover, many of these satellite systems are nowadays inactive, but their data can be still be used for research. Table 1 summarizes some of the general characteristics of several satellite data regarding spatial, spectral and temporal resolution. As indicated in Table 1, as a result of the space race, satellites have been able to monitor Earth since the 1960's. The Landsat program, which began in 1972 and continues to today, is considered a significant component of remote sensing applications in archaeology.

Prior to the Landsat program, satellite sensors such as CORONA and Zenit 2-8 sensors acquired only panchromatic photographs. These satellites were characterized by non-periodicity; therefore, some areas of archaeological interest may not have been photographed by these sensors. In contrast, the Landsat program has given further capabilities for research since the sensor is able to recover information in the visible, infrared and thermal part of the spectrum. Furthermore, the sun-synchronous orbit of the Landsat satellite enables researchers to study many archaeological sites and monuments in a systematic way. From the beginning of the Landsat program until the end of the century, new multispectral satellite sensors were launched from different countries, including the USA, USSR, France, and Japan, and the spatial resolution of the images was significantly improved. In 1999, the first high-resolution satellite imagery with a spatial resolution of less than 4m was available through the IKONOS space program. The IKONOS satellite was the first satellite operated by a private organization (Space Imaging). In 2000, NASA launched the first hyperspectral receiver, the EO-1 Hyperion, which had the ability to record electromagnetic radiation into 220 different spectral bands.

In the decade that followed, new satellites with higher spatial resolution were available to the scientific community and other countries became actively involved in space technology. Brief descriptions of different satellite sensors characteristics are highlighted in Table 1 and more specific information related to the most popular satellite platforms used in archaeolog- ical research are provided in the paragraphs below.

Landsat (MSS / TM / ETM +): The Landsat program was the result of the combined efforts of NASA and USGS to monitor Earth from space using remote sensing techniques. The first satellite launch was performed in 1972 (Landsat 1) and, since then, another 6 satellites were sent into orbit. According to Parcak (2009), the Landsat satellite program is the most well known satellite used for archaeological purposes due to its relative low cost, global coverage of the satellite data and access to archive data since the 1970's. Landsat satellite images cover an area of about 185 x 185 km. The multispectral bands of the sensor cover both the visible and infrared region of the spectrum while one sensor is able to produce thermal images. The panchromatic band of an ETM+ Landsat image has a spatial resolution of 15 m, while for the rest of the bands the resolution is set to 30 m with the exception of the thermal region (60 m). Landsat data can be obtained via FTP upon request from USGS (http://glovis.usgs.gov/).

CHRIS Proba: The Proba satellite belongs to a relatively new space program of the Europe- an Space Agency (ESA). The Compact High Resolution Imaging Spectrometer (CHRIS) sen- sor was launched on 2001 and provides hyperspectral images from 63 separate bands at a spatial resolution of 18 m. The objective of the CHRIS Proba is to evaluate new technologies for supporting future satellite sensors (experimental satellite) and to use the data for envi- ronmental purposes. The satellite data are acquired in HDF format after approval of ESA committee. A single satellite image covers an area of 13 x 13 km.

Table 1. List of available satellite sensors for archaeological purposes.

Satellite	Sensor	Acquisition period	Spatial resolutions		Spectral Resolution (nm) (only VIS-VNIR are listed)	Temporal Resolution
			Pan	VIS-NIR		
ALOS	PRISM	2006-Today	2.5	10	420 -890	46 days
CBERS	HRCC	2003-Today	20		450 - 890	26 days
CORONA		1960-1972	1.8 – 12		Panchromatic	
CARTOSAT-1		2005-Today	2.5		Panchromatic	116 days
EO-1	ALI	2000- Today	10	30	433-890	under req.
EO-1	Hyperion	2000-Today	10		356-996	under req.
FORMOSAT-2		2004-Today	2	8	450 -900	under req.
GeoEye-1		2008-Today	0.41	1.65	450 -920	under req.
IKONOS		1999-Today	1	4	450 -950	under req.
IRS	Cartosat-1 (IRS-P5)	2005-Today	2.5		Panchromatic	under req.

IRS	Cartosat-2B	2010-Today	1		Panchromatic	under req.
IRS	Resourcesat-1 (IRS-P6)	2003-Today	5.8	23.5	520 -860	under req.
IRS	Resourcesat-2	2011-Today	5.8	23.5	520 -860	under req.
IRS	1C / 1D	1996/7-Today	5.8	23.5	520 -860	under req.
KOMPSAT-2		2006-Today	1	4	450 -900	under req.
Kometa	KVR-1000	1981-2005	2-3		Panchromatic	
Kometa	TK-350	1981-2005	2-3		Panchromatic	
Landsat 4	MSS	1982-1993	60		520 - 900	
Landsat 5	TM	1984-Today	15	30	450 -900	16 days
Landsat 7	ETM+	1999-Today	15	30	450 -900	16 days
Orbview-3		2003-Today	1	4	450 -900	under req.
Pleiades-1		2011-Today	0.5	2	430-950	under req.
Proba	CHIRS	2001-Today	17-34		415-1050	under req.
QuickBird	-	2001-Today	0,60	2,4	450 -900	under req.
RapidEye		2008-Today	5		440 - 850	under req.
SPOT-1	HRV	1986-2003	10	20	500-890	
SPOT-2	HRV	1990-2009	10	20	500-890	
SPOT-3	HRV	1993-1996	10	20	500-890	
SPOT-4	HRVIR	1998-Today	10	20	500-890	under req.
SPOT-5	HRG	2002-Today	5	10	500-890	under req.
Terra	ASTER	1999-Today	15		520-860	under req.
Kometa	KVR-1000	1981-2005	2-3		Panchromatic	
TK-350			2-3		Panchromatic	
WorldView-1		2007-Today	0.5		Panchromatic	under req.
WorldView-2		2009-Today	0.5	1.8	400-1040	under req.
Zenit	2-8	1961-1994	15-2		Panchromatic	

EO-1 HYPERION: HYPERION was the first satellite of a new generation space program which was launched by NASA in 2000. The satellite's main objective was to collect experimental data for future receivers. The main feature of the HYEPRION satellite was the acquisition of hyperspectral data (a total of 220 separate bands) at a spectral range from 356 nm to 2577 nm. The spatial resolution of the data was 30 m. HYPERION data can be obtained via FTP upon request from USGS (http://glovis.usgs.gov/).

IKONOS: IKONOS is a commercial satellite with high spatial resolution. It was sent into orbit in 1999 and can provide images with spatial resolution up to 1m for panchromatic images and 4m in multispectral bands. The spectral resolution of the sensor extends from the visible to near infrared. Although IKONOS images are widely available to the research community, they are not recorded on a regular basis. The radiometric resolution of the satellite is 11 bit and a single image can cover an area of about 13 x 13 km. IKONOS satellite can provide stereo images in order to support the production of Digital Terrain Models and Surface Terrain Models (DEM, DSM). IKONOS data are available from GeoEye upon request (http://www.satimagingcorp.com/).

QuickBird: Quickbird is owned by the commercial satellite company DigitalGlobe and was sent into sun-synchronous orbit in 2001. The satellite is currently one of the few satellites with the highest spatial resolution (e.g. OrbView-2, OrbView-3, WorldView-1, WorldView-2 and GeoEye-1). The spatial resolution is up to 0.60 m in the panchromatic wavelength while multispectral bands are acquired at a resolution of 2.4 m. The spectral capacity is equivalent to the IKONOS satellite (visible and near infrared). Moreover, QuickBird images cover a ground area of 16.5 x 16.5 km. QuickBird data is available from DigitalGlobe after request (http://www.digitalglobe.com).

WorldView: WorldView satellite were launched in 2007 (WorldView -1) while a second sensor followed a few years later (WorldView-2). These sensors have a very high spatial resolution (0.5m). The WorldView-2 sensor provides a high resolution panchromatic band and eight Multispectral bands; four standard colors (red, green, blue, and near-infrared) and four new bands (coastal, yellow, red edge, and near-infrared). WorldView data is available from DigitalGlobe upon request (http://www.digitalglobe.com).

GeoEye-1: GeoEye is the latest high spatial resolution satellite that was sent into space (2008). The spatial resolution of the satellite is 0.41 m and 1.65 m (panchromatic / multispectral bands). The spectral resolution is limited to visible and near infrared wavelength. A Ge- oEye-1 image covers an area of 15 x 15 km.

CORONA: From 1960 until 1972, the CORONA satellite acquired over 860,000 panchromatic images for US Intelligence. The photographic capsule from the spy satellite was dropped to earth with the help of parachute and then was collected by a special aircraft (Figure 1). The CORONA images were declassified in 1995, and are now available in digital form upon request.

Remote sensing has been able to assist archaeological research in several ways during the past years, including detection of subsurface remains, monitoring archaeological sites and monuments, archaeolandscapes studies, etc. The next section presents recent developments and applications of several remote sensing techniques for supporting archaeological re- search. The section includes detection of subsurface remains at the Thessalian plain based on both satellite and ground spectroradiometric measurements. Moreover, remote sensing and GIS analysis as means for monitoring purposes in the area of Cyprus are also examined. Geophysical surveys from various archaeological sites are also presented as well as the re- sults of a study aiming to analyse the impact of atmospheric pollution on archaeological sites. The section ends with discussion of low-altitude airborne systems, as well as 3D laser scanner documentation of cultural heritage site.

Figure 1. Film capsule of the CORONA satellite collected from aircrafts. (Photos from Wikipedia and CSNR collection)

3. Monitoring archaeological sites using satellite remote sensing and GIS analysis

In many areas of the world, cultural heritage sites and visible monuments are monitored mostly with on-site observations, including data collection, periodic observations for archaeological sites and multi-analysis risk assessments. In this way, on-site observations are time consuming and not cost-effective.

Hadjimitsis et al., (2011) highlighted the beneficial integrated use of satellite remote sensing with GIS for exploring the natural and anthropogenic hazard risk of the most significant cultural heritage sites in Cyprus. In order to proceed to overall risk and vulnerability assessment of the archaeological sites in Cyprus due to anthropogenic and natural impact, a risk index was attributed to each different factor such as urban activity, minimum distance of urban activity in the vicinity of an archaeological site, seismic PGA and air pollution impact. They found that, concerning the seismic risk assessment, that significant monuments are located within the spatial limits of the most seismic prone areas in Cyprus. Additionally, regarding sea erosion, the study proved that 50% of the sites examined in the study, are within a distance of only 500 m away from the coastline making them vulnerable to related coastal hazards such as sea water erosion. The creation of buffer zones in GIS environment around CH sites explored the significant problem of extensive urbanization in the vicinity of cultural heritage sites. Almost 50% of the CH sites are under severe urban pressure and a percentage of 37.5% of the sites are within a radius of 500m from the urban centers. In similar studies, Carlon et al., (2002) and (Alexakis and Sarris, 2010) used both anthropogenic and natural factors to cre- ate a risk assessment model concerning archaeological monuments in Venice and West- ern Crete respectively. Moreover, Urhus et al (2006) emphasized the human driven agents, such as camping, hunting and woodcutting, for assessing the modern threats to heritage resources and Lanza (2003) addressed the potential threat that is posed at the historical center of Genoa in the case of failure of the urban drainage system.

This section presents the contribution of remote sensing for monitoring the surroundings of archaeological sites in order the managing authorities or governmental related bodies to be able to conduct a risk assessment analysis of cultural heritage sites in Cyprus. Figure 2 presents some of the most indicative threat parameters. Special attention in this section is

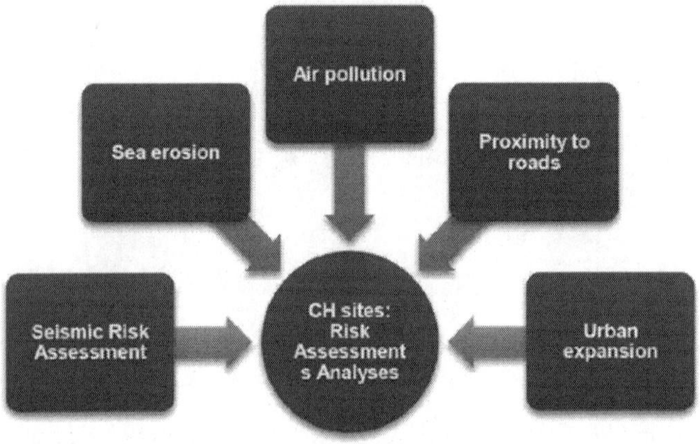

Figure 2. Risk assessment analysis for cultural heritage sites (Hadjimitsis et al., 2011)

given to urban expansion during the past 50 years. Anthropogenic factors, such as urban ex- pansion and air pollution contribute significantly to the destruction of cultural heritage sites. Remote sensing and GIS provide synoptic views of cultural heritage sites which enable policy makers to make appropriate decisions regarding the preservation of cultural heritage sites.

3.1 Urban expansion and other hazards as a threat to archaeological sites

In order to study and map urban expansion, a number of significant archaeological sites of Cyprus were examined. These cultural heritage sites are located in the southern coastal part of the island (from west to east): *Tombs of the Kings, Nea Paphos, Palaepaphos (Old Paphos),* and *Amathus.* Urban expansion was monitored with the extensive use of time-series multispectral and aerial dataset. All images were both geometrically and radiometric corrected in ERDAS Imagine 9.3 software. Moreover, atmospheric correction was also performed based on the Darkest Pixel algorithm (see Hadjimitsis et al., 2009, 2002; Agapiou et al., 2011). Post-processing techniques included histogram enhancement, computa- tion of vegetation indices, band ratios, principal component analysis and photo-interpretation of the results.

The results showed a dramatic increase in urban expansion of main cities of Cyprus (Limassol and Paphos) during the last 50 years. For example, in the case of the *Palaepaphos* site (Figure 3), the entire east area of Kouklia village (*Palaepaphos*) is still undeveloped, while at the west area the urban expansion has been increase dramatically (Agapiou et al., 2010a).

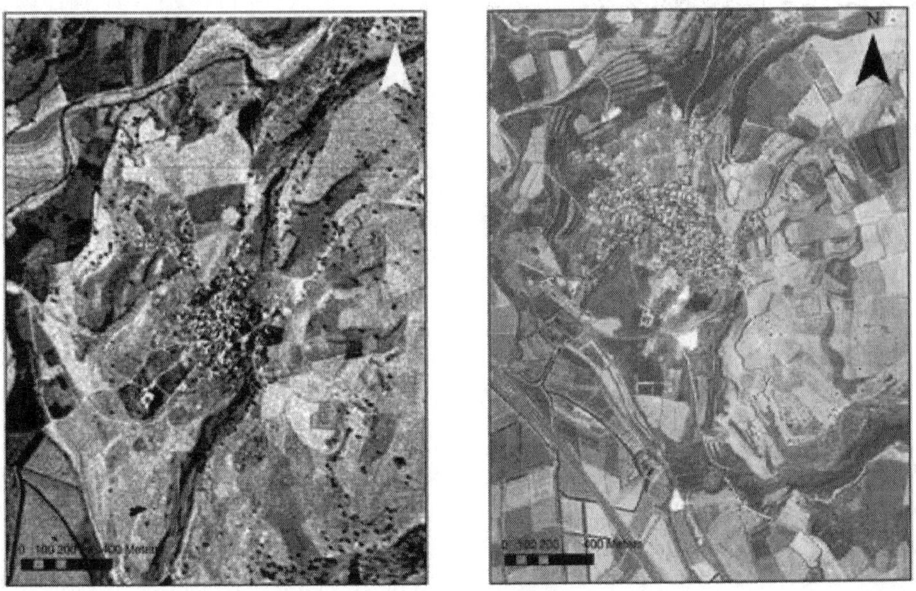

Figure 3. Palaepaphos archaeological site in 1963 CORONA image (left) and 2004 QuickBird image (right) (Hadjimitsis et al., 2010)

Urban sprawl has been recorded also in the broader area of Paphos during the last decades. Extensive construction and building development has taken place and areas with significant archaeological interest are now affected from urban expansion. Thus, the land use and land cover region of the area was examined to monitor and map the size of urban expansion in the vicinity of the archaeological sites of Tombs of the Kings and Nea Paphos during the last half century. Aerial photos of the study area, acquired in 1963 and 2008 were provided from the Department of Lands and Surveys of Cyprus. Initially, aerial photos were georeferenced in a GIS environment with the use of ground control points (GCP's). The digitization of all the buildings in the broader area of Nea Paphos and Tombs of the Kings was performed for both time periods. Their direct comparison enabled the researchers to map the extent of urban development during the last years and revealed the impact of urbanization on the preservation of archaeological sites (Figure 4).

Figure 4. Urban expansion near the archaeological sites of Nea Paphos and Tombs of the Kings during the last 50 years (3D view).

CORONA satellite images have also indicated the growth of the urban activity around the *Amathus* archaeological site, including the highway that passes 100 m north of the site (see Figure 5) (Hadjimitsis et al., 2010). Several satellite images were used to examine the threat of urban expansion around the *Amathus* archaeological site located just east from the out- skirts of the city (Figure 6). The dataset includes Landsat TM/ETM+ images from 1987 until 2009. As shown in Figure 6, urban expansion is clearly observed though interpretation of the images.

It is very important for researchers to understand the dramatic changes that have occurred due to human activity during the last decades. Figure 7 highlights the potential risk of the archaeological sites due to urban expansion of the city of Limassol. Using archive satellite images, the researchers can map this expansion with great detail and accuracy based on classification techniques.

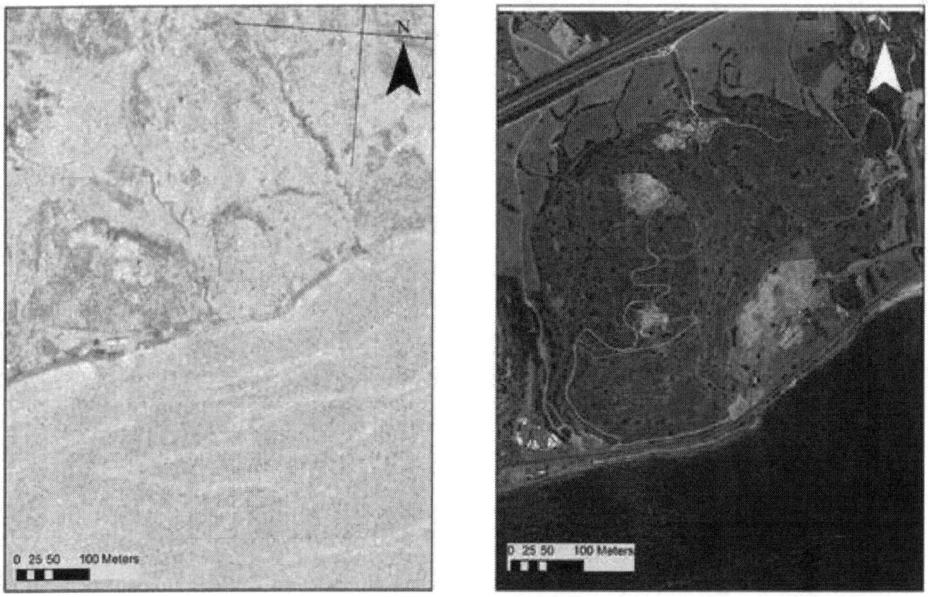

Figure 5. Amathus archaeological site in 1963 CORONA image (left) and 2010 Google (right).

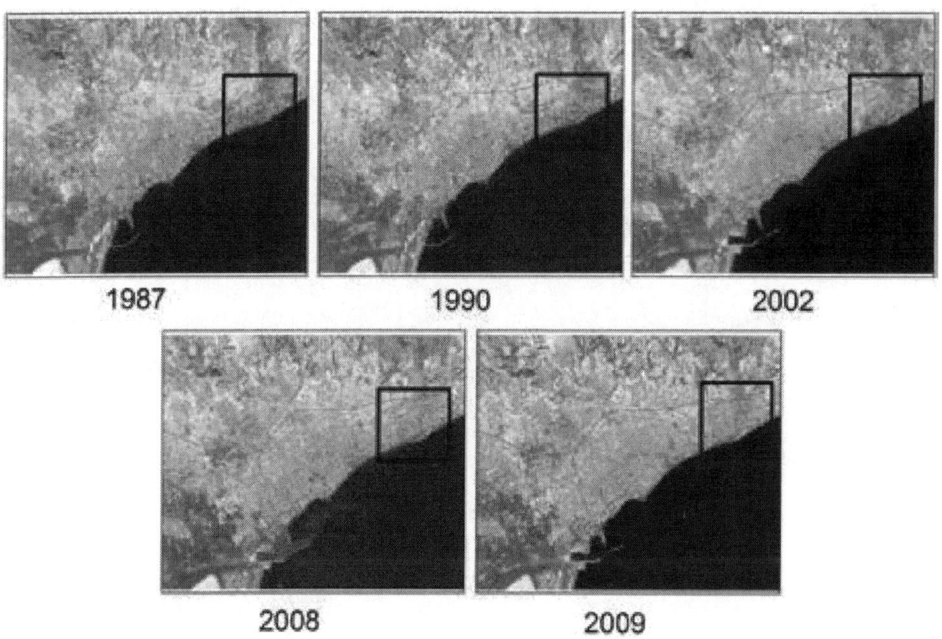

Figure 6. Landsat images used for mapping the urban expansion of Limassol town during the last 30 years. Amathus archaeological site is indicated in a square.

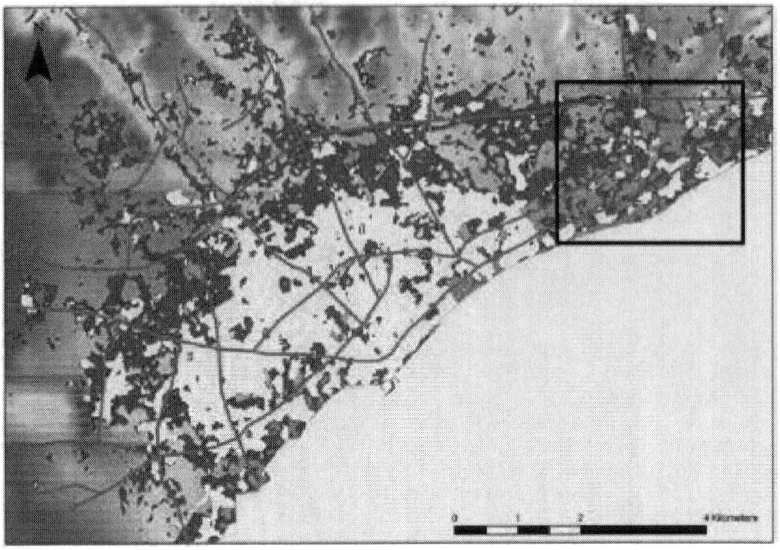

Figure 7. Urban areas of Limassol town in 1987 (red) and in 2009 (pink). The Amathus archaeological site is indicated in a square.

Vegetation indices are also a key parameters that can be used for monitoring dramatic land use changes over time (e.g. urban activities). The Normalized Difference Vegetation Index (NDVI, with range -1 to +1) was applied to the entire dataset (Figure 8). High values of NDVI (indicated with green in Figure 8) are present vegetated areas while low NDVI values (indicated with yellow) are recorded for areas with no vegetation. Since NDVI values may vary throughout time due to the physical phenological changes of the plants, similar periods of Landsat images were examined.

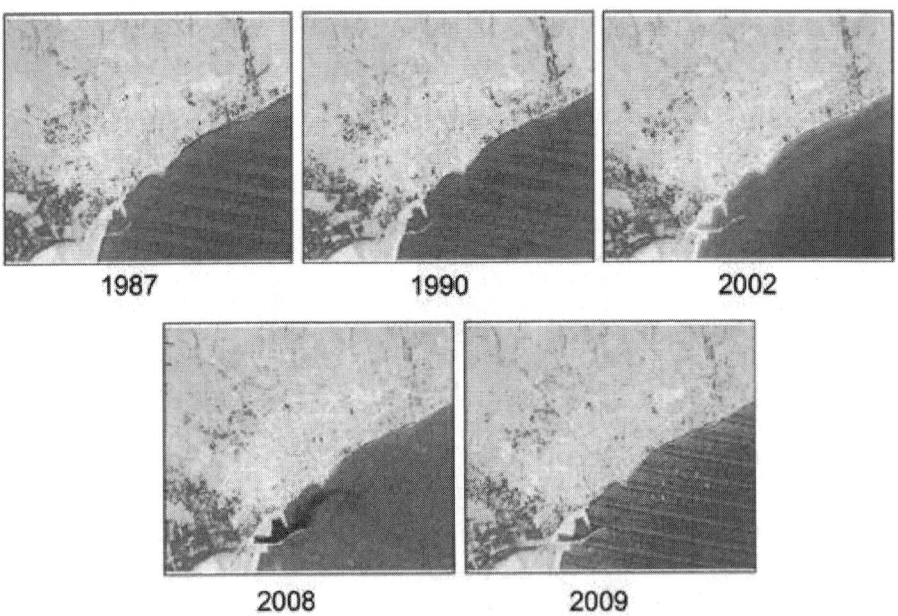

Figure 8. NDVI maps produced from Landsat dataset.

NDVI values were used along with classifications results in order to record NDVI differences in urban classified areas. Figure 9 demonstrates the results of the NDVI difference for the period 1987-2009. Although many areas have indicated no dramatic changes, some other areas represented in yellow and red colour (Figure 9) indicate dramatic transformation of the initial landscape. Indeed, such changes have been recorded in a very close proximity of the archaeological site of *Amathus* (see Figure 9 in black square).

Further anthropogenic and natural hazards (e.g. landslides; sea erosion; earthquakes etc) can be monitored in a systematic basis using remote sensing data and GIS spatial analysis. Different studies (Hadjimitsis et al., 2010; 2011) have shown the potential of using such methodologies for cultural heritage risk assessment.

Contemporary technological means such as GIS and satellite remote sensing provide efficient and detailed maps of the region of CH sites in the island of Cyprus. This specific study revealed the different kinds of natural and anthropogenic hazards that threaten the preservation of valuable CH sites.

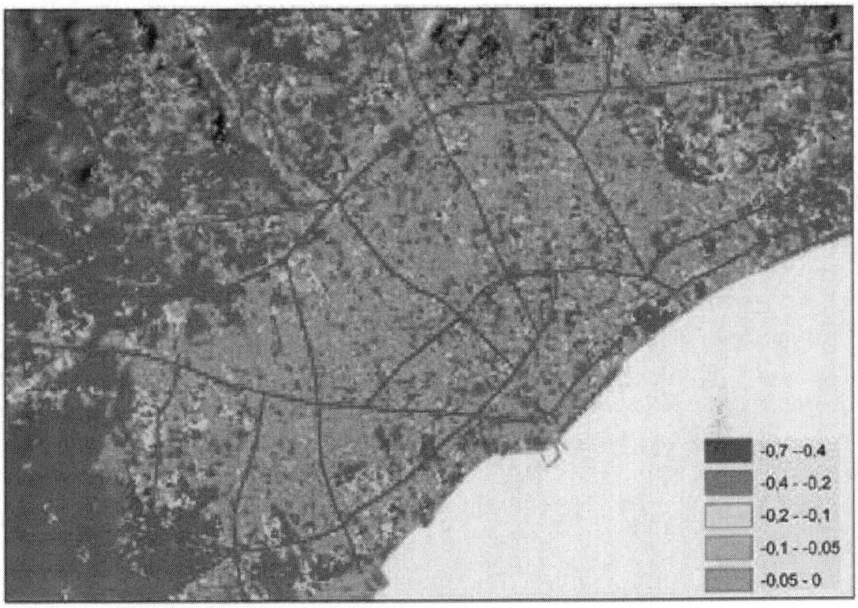

Figure 9. NDVI difference from 1987 until 2009.

3.2 Monitoring air quality in the vicinity of archaeological sites based on satellite and ground measurements

Although cultural heritage sites are documented and preserved, there has been limited monitoring and documentation of how cultural heritage sites are affected by air pollution. Themistocleous et al., (2012a) introduced a new approach for monitoring air pollution near cultural heritage sites. By using a variety of tools, including satellite images, sunphotometers, PM_{10} monitors, and laser scanners, the level of air pollution and its effect on cultural heritage sites can be determined. The cultural heritage sites were documented, and using GIS tool, any significant areas of air pollution, including urban areas, industrial areas, and roads were determined. The algorithm proposed by Themistocleous (2011) was applied to retrieve the aerosol optical thickness (AOT) from Landsat TM/ETM+ satellite images in order also to cross-validate the AOT values found from MODIS and sun-photometers.

Spectral variations recorded by satellite sensors are indicators of aerosol particles and, therefore, air pollution. The key parameter for assessing atmospheric pollution in air pollution studies is the aerosol optical thickness. Aerosol optical thickness (AOT) is a measure of aerosol loading in the atmosphere (Retalis et al., 2010). High AOT values suggest high concentration of aerosols, and therefore air pollution (Retalis et al, 2010). The use of earth observation is based on the monitoring and determination of AOT either direct or indirect as tool for assessing and measure air pollution. Several studies have shown that satellite data can be used to monitor air pollution and air pollution effects. Tømmervik et al., (1995) compared vegetation cover maps and air pollution emissions data over a 15 year period and found major changes in the environment as a result of high air pollution values. Nisantzi et al., (2011) used MODIS satellite data to analyse the relationship between the aerosol optical thickness (AOT) and the PM_{10} as indicators of pollution. Satellite remote sensing can be used to assist in air quality monitoring and identify the need to protect cultural heritage in urban areas from air pollution (Hadjimitsis et al., 2002; Kaufman et al, 1990; Retalis, 1998; Retalis et al., 1999). Pollution not only deteriorates cultural heritage sites but may also cause irreversible damage that prevents the proper salvation of the monument (Skoulikides, 2000). Therefore, improving air quality is critical for the preservation and maintenance of cultural heritage sites.

The study area was the Limassol Castle, located in the center of Limassol, Cyprus. The study utilized a variety of remote sensing tools to measure air pollution. Landsat TM/ETM+ and MODIS satellite images, as well as the GER 1500 spectro-radiometer, were used to directly or indirectly retrieve AOT, as were ground measurements using the Microtops II handheld sun- photometer and the Cimel sun-photometer located at the Cyprus University of Technology, which is part of the AERONET program. Air particles'

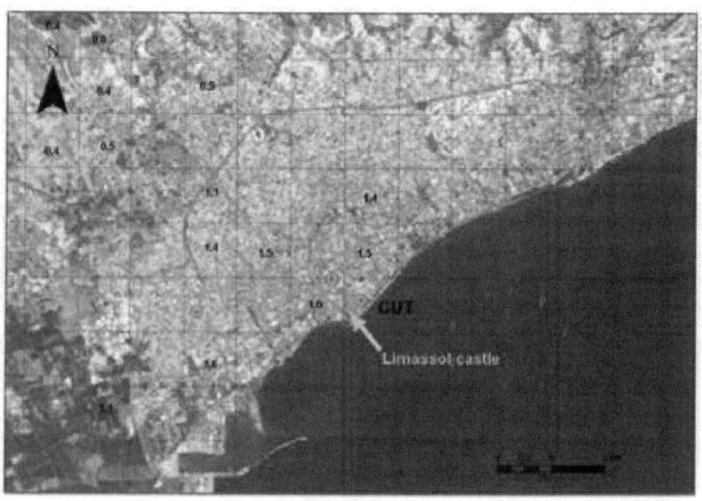

Figure 10. AOT levels in the Limassol area. High AOT levels are noted in the area near the Limassol Castle.

measurements were correlated to the AOT levels to verify the level of pollution. Last, visual observation of the Limassol Castle identified the damage caused by air pollution and laser scanning to document and monitor the damage was conducted. Results from satellite remote sensing identified that the centre of Li- massol contains high levels of air pollution, with values of AOT higher than other surrounding areas. Determination of AOT measurements using MODIS and Landsat satellite images found that the centre of Limassol, where the Limassol Castle is located, experiences the highest level of AOT values (Figure 10). A PM_{10} /$PM_{2.5}$ in situ measurement campaign in the area of the Li- massol Castle found that for the majority of the time periods, the PM_{10} readings exceeded the limit value (50 µg/m3), indicating a high level of air pollution in the area.

A similar approach was followed for the Paphos town using daily MODIS AOT data. The results have shown that 54% of the measurements for air quality was above the threshold of AOT 300 (AOT 0.300) (see Figure 11). This analysis suggest that cultural heritage sites near the Paphos town (e.g. *Nea Paphos, Tombs of the Kings* etc) are exposed to air pollutants half the time.

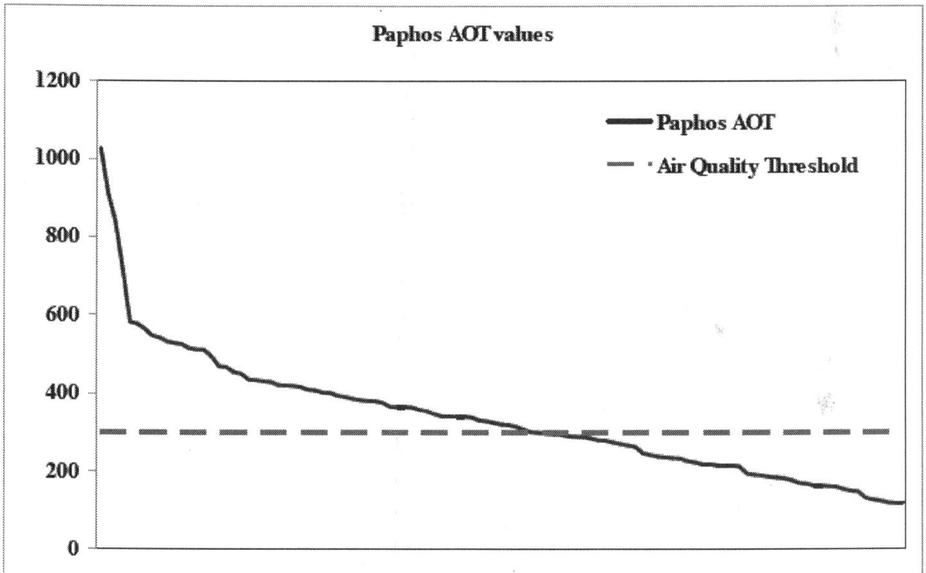

Figure 11. Paphos AOT values (sample = 109 measurements) in blue. In red circle is the threshold air quality limit of 300 (AOT 0.300). In the y-axis, AOT value is multiplied by 1000 (to match MODIS data) (Themistocleous et al., 2012a).

4. Detection of archaeological sites based on remote sensing techniques

Several Neolithic settlements ("magoules") are located in the Thessalian plain in central Greece. These sites are typically found as low hills raised up to 5-10 m. Alexakis et al., (2009; 2011) has recently shown that the detection of several unknown sites is possible based on remote sensing and GIS analysis. The study aimed to combine several types of remote sensing data (e.g. Landsat TM/ETM+, ASTER, Hyperion, IKONOS) and DEM in order to improve the detection of these subsurface remains (Figure 12). The satellite data were statistically analyzed, together with other environmental parameters, to examine any kind of correlation between environmental, archaeological and satellite data. Moreover, different methods were compared for the detection of Neolithic settlements. The results of the study suggested that the complementary use of different imagery can provide more satisfactory results.

Further to the Alexakis study, Agapiou et al., (2012a) argued that the detection of the settlements is possible based on ground spectroradiometric measurements. Several spectroradio- metric measurements have indicated that each magoula has its own spectral characteristics related to its own morphological characteristics. The study has found that the highest peak of the magoula tends to give high NDVI and SR values (similar to the flat – healthy regions) while the slope of the magoula has lowest NDVI and SR values (and for the other indices as well). The extraction of each magoula requires further analysis and enhancement techniques in cases where the spatial resolution of the satellite image used is low. Local histogram en- hancements can identify magoules as a small difference of NDVI values at the same parcel (Figure 13).

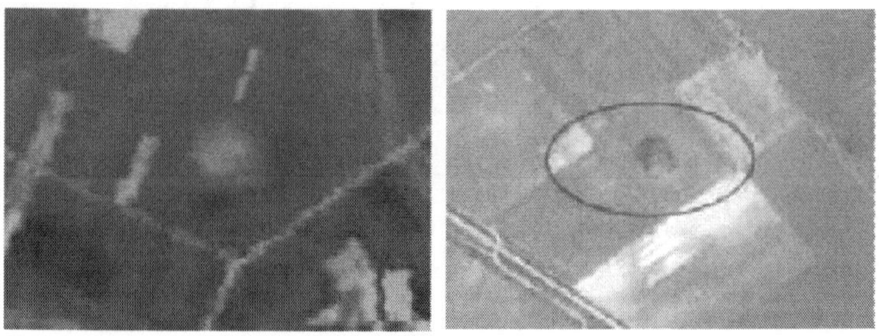

Figure 12. Magoula Neraida using ASTER image (left). Magoula Melissa 1 using IKONOS image (RGB - 321) (right).

Figure 13. NDVI results for Prodromos II site (in green circle). (a) Raw satellite image without any radiometric enhance- ments, (b) satellite image with a linear max-min enhancement applied to all image, (c) max-min enhancement applied to the area around Prodromos II and (d) modified max-min enhancement applied to the area

around Prodromos II. The magoula is indicated with the red arrow (Agapiou et al., 2012c).

Similar results were found following the application of the Tasselled Cap algorithm (Figure 14 to a series of Landsat TM/ETM+ multispectral images. The Tasselled Cap transformation is used to enhance spectral information for Landsat images, and it was specially developed for vegetation studies. The first three bands of the Tasseled Cap algorithm result are characterized as follow: band 1: brightness (measure of soil); band 2: greenness (measure of vegetation); band 3: wetness (interrelationship of soil and canopy moisture).

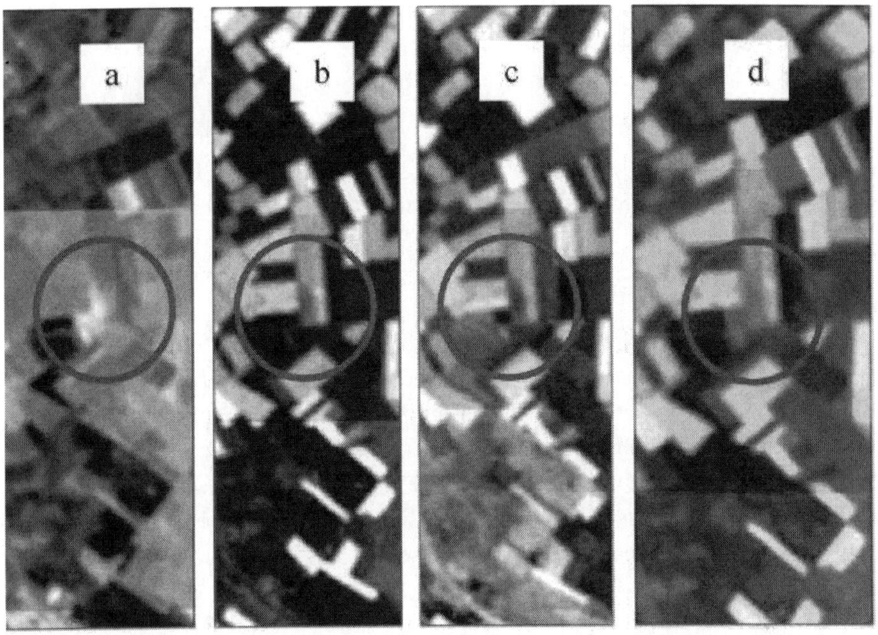

Figure 14. Tasseled Cap results for Nikaia 16 site (in red circle), (a) Brightness, (b) greenness, (c) wetness and (d) RGB of the first three components of the T-K algorithm (Agapiou et al., 2012c).

Phenological studies of crops for the detection of buried archaeological remains were also evaluated (Agapiou et al., 2012b) It was found that the phenological cycle of crops for 'archaeological' and 'non archaeological areas' can be used as a "remote" approach in order to locate buried architecture remains. In Figure 15, the phenological cycle of an archaeological site (*Almyros II*) and the phenological cycle of a healthy site (Site 3) are examined. A small NDVI difference is evident (Case A, Figure 15) which is associated with buried archaeological remains. This is due to the fact that soil over the archaeological remains seems to have a different moisture content compared to their surroundings. Therefore, although there exist similar climate characteristics and crop cultivation techniques, there is a difference in amplitude of the NDVI cycle of the archaeological and non- archaeological areas.

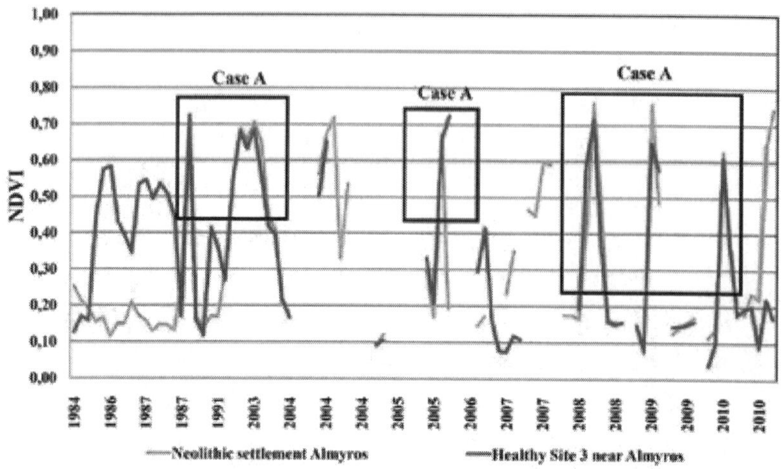

Figure 15. Phenological cycle of the Neolithic settlement (solid line) and the healthy site 3 (dashed line) (Agapiou et al., 2012b)

5. Documentation of cultural heritage sites using remote sensing techniques, GIS and laser scanning

Contemporary techniques and methods such as computer graphics, virtual reality, multime- dia technology, and information technology can be integrated in Web GIS technologies, in order to act as a uniform digital tool for documentation, protection and preservation of cul- tural heritage (Agapiou et al., 2010c; Hadjimitsis et al., 2006). In order to document and map known archaeological sites and monuments, several techniques may be used, including la- ser scanning, 3D modelling and GIS. In this section, applications from several monuments in Cyprus are presented.

5.1. Integrated use of GIS and remote sensing: a pilot application at the archaeological sites of Paphos

Local cadastral maps were used to support the documentation of cultural heritage sites in the Paphos district, SW Cyprus. In general, each monument may be located in a different sheet /plan; therefore, spatial analysis from such data is a very difficult task.

In order to overcome such limitations, a GIS geodatabase was developed using the ArcGIS 10 software. A GIS system is a computer system (software) that collects, stores, manages, an- alyzes and visualizes spatial information and upgrades to other information systems. There- fore, GIS can be used as a tool for modelling and analysis of complex research and as a system that supports decision making. Important advantages of GIS include: (a) The data can be stored in a small digital space, (b) Both the storage and the recovery can be achieved with lower costs than traditional ways, (c) Analysis can be carried out much faster, (d) GIS allow synthetic analysis of data without any particular problems and (e) GIS offers the digi- tal environment for an integrated process, where the collection, analysis and decision proc- ess are in a continuous flow.

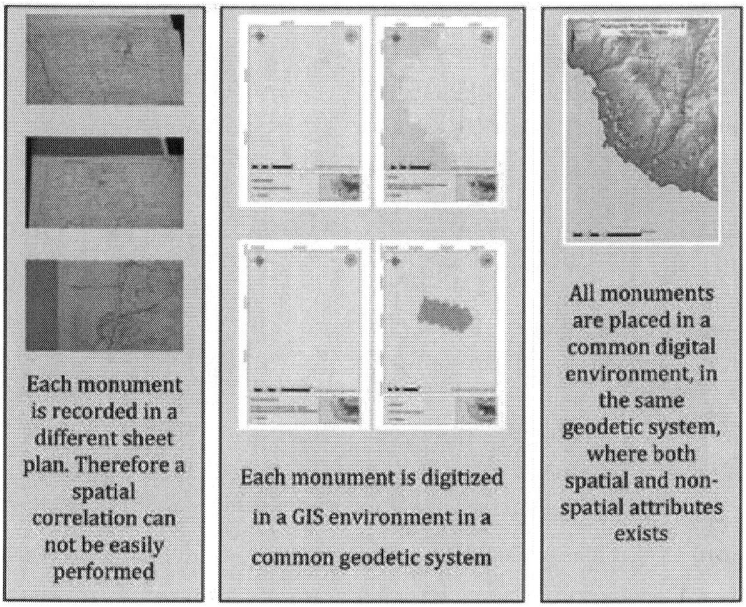

Figure 16. Methodology of mapping the archaeological sites

The most important advantage of the GIS environment is that it can connect both spatial in- formation (e.g. place, coordinates) along with a-spatial (non-spatial) information (e.g. type of the monument, chronology etc). In this way, further spatial analysis can be performed (Figure 16).

For each monument listed by the Department of Antiquities of Cyprus (200 monuments belonging to the Paphos district), the relative sheet plan was found and digitized. All monuments were georeferenced in a common geodetic system (WGS 84, 36N) (Figure 17). The overall map created (Figure 18), can assist risk assessment analysis. Such kind of an integrated CHM/GIS system has been recently implemented to be used for the efficient manipulation of information regarding the ancient monuments and movable antiquities of Cyprus (Kydonakis et al 2012).

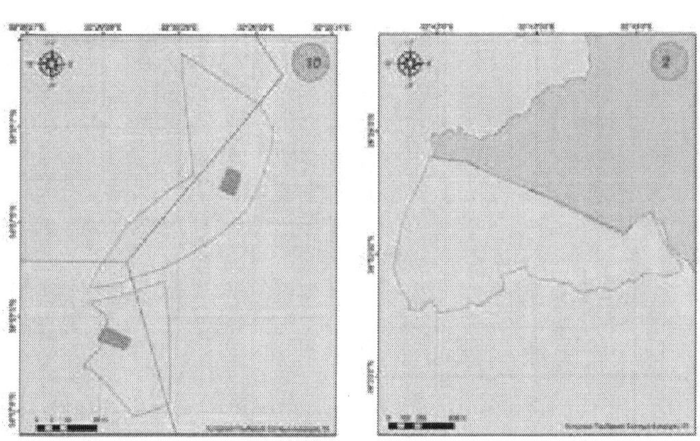

Figure 17. Example of the mapping procedure using the GIS software.

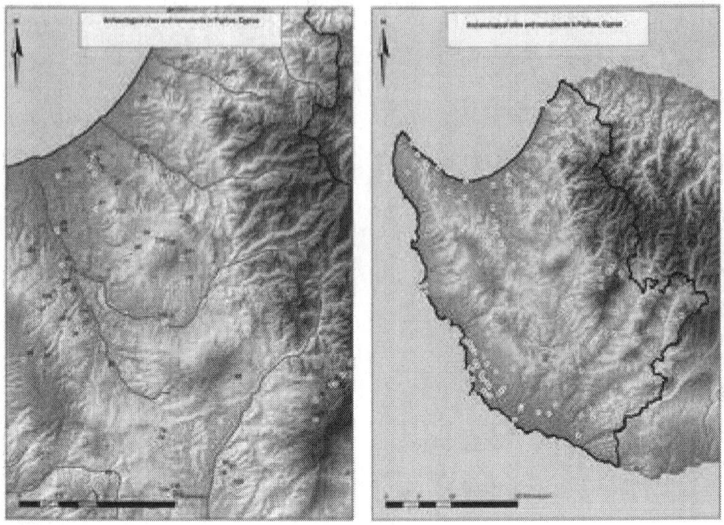

Figure 18. Archaeological sites and monuments of the Paphos District.

5.2. Terrestrial laser scanning for documentation, reconstruction and cultural heritage structural integrity

Due to their high data acquisition rate, relatively high accuracy and high spatial data density, terrestrial laser scanners are increasingly being used for cultural heritage recording, architectural documentation studies, research of cultural heritage with photogrammetric methods and engineering applications that demand high spatial resolution. Terrestrial laser scanning process can be considered as a part of remote sensing methods. In this section, the results from three different cases studies are presented: *Saint Theodore, Tomb I* at the *Tombs of the Kings* and the *Church of Kyrikos and Ioulitis*

For the documentation of the church of *Saint Theodore* in Idalion village, central Cyprus, the 3D laser scanner Leica C10 was used (Figure 19). Pre-processing of the point clouds was performed at the Cyclone software. The latest includes the noise removal of the initial point clouds and the registration using scan targets (Agapiou et al., 2010b).

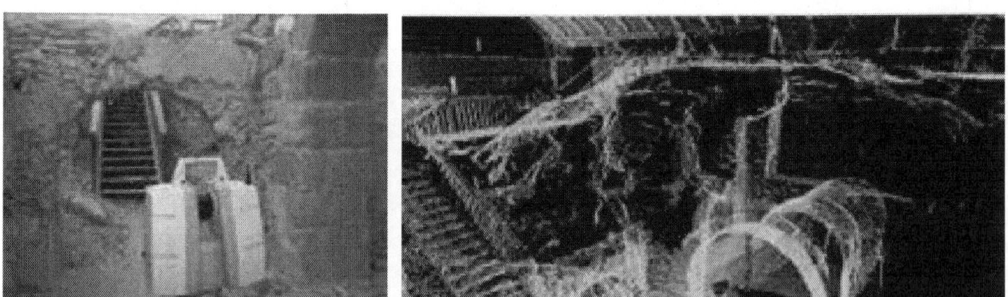

Figure 19. Data collection from the church of *Saint Theodore* in Idalion village (left). Registration of the point clouds for *Saint Theodore* in Idalion village. All point clouds are transformed into one coordinate system (right) (Agapiou et al., 2010b).

A single scan station was also used for the interior of the *Tomb I*, located at the *Tombs of the Kings*, archaeological site. The data were then processed at the Cyclone software. The initial point cloud of the Tomb I was further analysed and a 3D mesh was finally created (Figure 20). Using the 3D mesh several sections can be drawn in order to study in detail the architecture of *Tomb I*.

The third example is the *Saint Kirikos and Ioulitis* church. Specific laser scans were acquired from the exterior and the interior of the church. The use of laser scanner can provide accurate geometric documentation of such buildings through time and monitor them. One such example is the crack presented in the background of fresco of Christ in the church of Saint Kirikos and Ioulitis (Figure 21). Repeated accurate measurements of the order of magnitude of a few mm can identify if the crack is gradually increasing in size.

The combination of 3D model and WebGIS applications was also presented by Agapiou et al., (2010c). The "Digital Atlas of Byzantine and Post Byzantines churches" application consists of a WebGIS tool, using the ArcGIS Server software. The WebGIS includes a detail 3D reconstruction of the surrounding area of the monuments using grayscale high resolution orthophotos, a digital elevation model (DEM) of a high accuracy of (± 2m) and 3D digital "light" models of the monuments, produced in Google SketchUp software, after applying topometric methods for measurements. Moreover, the application includes non-spatial information about the monuments, such as relevant bibliography, photos of the interior and exterior of the monuments and also audiovisual data. Finally, this digital tool provides to the end-users a brief, time-stamped, historical background information about the Byzantine and post-Byzantine monuments of central Cyprus (www.byzantine-cyprus.com).

Figure 20. Mesh documentation of the interior of the Tomb I, Tombs of the Kings archaeological site.

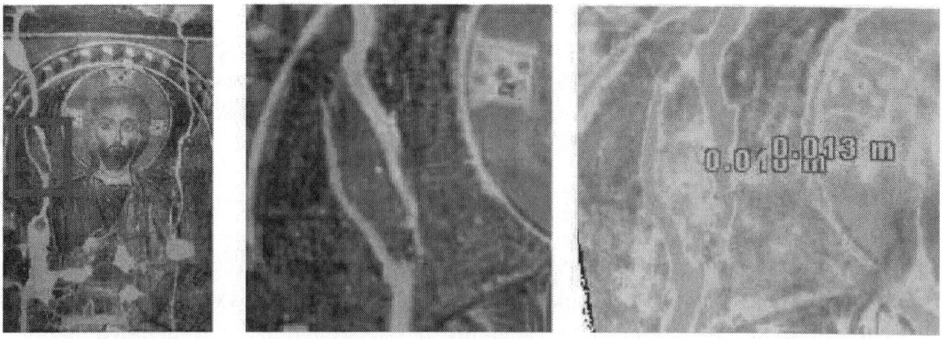

Figure 21. Monitoring the crack (see square in the first image from the left) of the background of the fresco at Saint Kirikos and Ioulitis through Laser Scanners (Agapiou et al., 2010b).

Figure 22. Models for Byzantine and Post Byzantine churches of Cyprus using topometric measurements and GIS tools (Agapiou et al., 2010c).

Moreover, laser scanners can be used for monitoring purposes as shown by Themistocleous et al., (2012a). In order to monitor the effects of air pollution, the Limassol Castle is being documented every year with the 3D laser scanner. Areas of the castle which show deterioration on the 3D laser scanner will have samples taken to determine the chemical analysis of the surface to establish if the deterioration was caused by air pollution or natural causes. Photographs of the castle were also taken and applied to the 3D laser scanned point cloud. A direct visual comparison between the intensity of the laser scanner and close range photo- graphs of the cracks in the Limassol Castle indicate that observation of intensity values can indicate the presence -or not- of possible cracks in the monument. (Figures 23 and 24). Similar conclusions can be drawn when laser scanner intensity is compared with ultrasonic measurements.

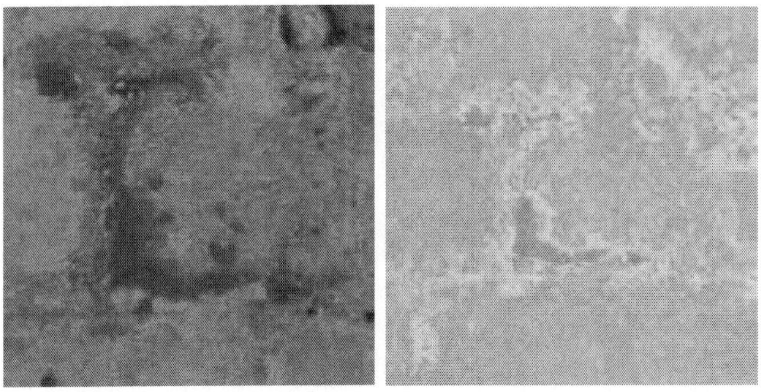

Figure 23. Visual comparison of the laser intensity and close range photographs near a crack

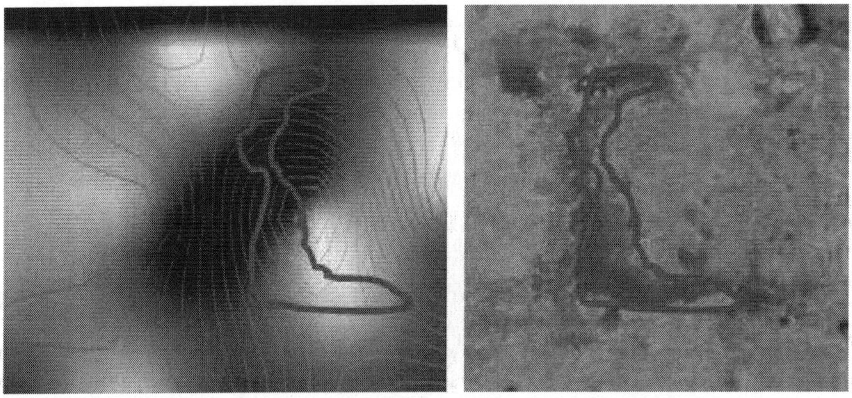

Figure 24. Visual comparison of the ultrasonic measurements and close range photographs. The polygons are drawn as common areas for each set of figures.

6. Geophysical prospection techniques: From mapping to CRM

In terms of ground based remote sensing, there is a wide range of surveying techniques that are focus targeted towards the shallow or medium mapping of the subsurface antiquities or even of the deeper geological layers that may have covered the cultural strata. The various methods, including magnetometry, soil resistance or electromagnetic methods (EM), ground penetrating radar (GPR), and seismic, are based on the measurement of different physical quantities and the complementary application of them (the manifold approach) produces datasets that can match each other and maximize the information content of the geophysical interpretation (Sarris, 2012). Depending on the method and the configuration of the techniques, it is also possible to have different penetration depths and operation in diverse environmental settings (rural or urban) to address a various topics related to the mapping of archaeological sites and archaeo-environment, the preservation of monuments, e.t.c. Geophysical approaches can be applied in planned excavations, rescue archaeology, archaeolandscape studies, building conservation and cultural resources management (Sarris & Jones 2000).

In general, magnetic techniques using the measurement of the total geo-magnetic field intensity or of the gradient of it or one of its components can be helpful in identifying architectural relics or residues of habitation and workshop activities. Magnetometry techniques have been successfully used to map the relics of settlements and reveal the town planning system. Mud brick foundations of Late Neolithic houses together with pits and other details were recorded around the tell of *Sceghalom-Kovácshalom* in E. Hungary. The organic material gathered in the pits was responsible for the enhancement of the magnetic susceptibility, resulting in the good registration of the pits from the measurements of the vertical magnetic gradient. Even stronger was the magnetic signature of the foundations of the fired daub foundations and walls of the farmsteads that were recorded as thermal targets, but which at the same time were not able to register to the GPR measurements due to the high conductivity of the soils (Monahan & Sarris 2011, Sarris, 2012) (Figure 25). The same type of thermal signature is shown in the investigation of workshops and kilns belonging to different chronological periods. In other cases, such as in *Sikyon*, Peloponnese (S. Greece), the difference of the construction materials of the structural remains of the Hellenistic/Roman city in terms of the magnetic minerals they contained was responsible for providing an accurate plan of the ancient city. Due to the soil conditions and the preservation of the site, the magnetometry survey specified the street layout and the city quarters, tracing numerous monuments inside and outside the agora limits, including temples, porticoes, a basilica, street lines, houses and industrial installations (Sarris et al., 2009; Gourley et al., 2008).

Similar is the operation of the EM and soil resistance methods, which, together with the GPR, are considered ideal to resolve features related to structural remains, champers, voids and tombs. These methods are considered to be active measuring techniques. The particular methodology has been used successfully in resolving the foundations of buildings, road networks, and funeral residues. Of particular interest is their ability to operate in different frequencies (EM and GPR) or configurations (soil resistance) allowing a larger or smaller penetration depth. In this way, it is possible to provide valuable information regarding the subsurface stratigraphy. For example, the decrease of the GPR antenna frequency can provide a larger penetration to the soil strata. In addition, the multiple reflections of the GPR electromagnetic signals originating from adjacent (usually parallel) transect can create images of the subsurface layers (of various widths) by increasing depth (depth slices) (Figure 25). In a similar way, vertical electric soundings measure re-sistivity variations with depth by increasing gradually current electrode separation while the center of the electrode configuration, remains stationary. Based on the same principle, the electrical resistivity tomography provides information for both the lateral and vertical variations in the resistivity of the soil and, based on 2D or 3D inversion algorithms; it can produce a 3D reconstruction model of the subsurface (Papadopoulos et al., 2011, Sarris 2008).

The use of the EM, electrical resistivity tomography (ERT) and seismic techniques is more appropriate for the deeper mapping and their employment is usually applied in ar-chaeolandscape studies. This was the case of *Priniatikos Pyrgos*, where the integrated application of ERT and seismic tomography techniques processed by 3D inversion algorithms were capable to contribute to the archaeoenvironmental reconstruction of the *Priniatikos Pyrgos* at Istron, E. Crete, providing indications regarding the ancient harbor of the nearby settlement (Shahrukh et al 2012). The particular methods were the only solution to provide information about the deposits that exist in the coastal area of *Priniati- kos Pyrgos*: carstic formations of medium to high permeability and alluvium deposits of variable permeability, probably originating by past landslide episodes and periodic flooding of the Istron River, have covered the ancient harbour at depths varying from 20-40m below the current surface. Similarly, electromagnetic and soil resistance measurements re- vealed the movement of the older Istron River branches, which appeared to be directed to the sea from both sides of the settlement, leaving probably a small path to the main- land from the SW direction. The above results were also supported by the sedimentological analyses and OSL dating of cores taken from the region and the use of geophysical techniques in the study of the dynamics of the landscape evolution (Sarris et al 2012) (Figure 26).

GPR and soil resistance techniques (including ERT) also can be used in an urbanized context in contrast to the rest of the geophysical approaches (Sarris 2008; Linford 2006). Due to a high level of ambient noise from the background anthropogenic activities and the high dis- turbance of the upper soil layers, the particular techniques can be adapted to resolve a num- ber of issues in question (Sarris & Papadopoulos 2011; Papadopoulos et al., 2009). Thus, the above methodology can be used during the course of private construction activities but also for even larger civil construction works that can deal with highways, squares, pedestrian roads, etc. In a number of instances they can even be applied within historical structures and monuments to conclude on the integrity status of the monuments. The geophysical techni- ques can also contribute to a more generalized risk assessment model, since it can provide information for the tectonic regime and the classification of geological strata either in terms of their resistivity (ERT), velocity of propagation of acoustical waves (seismic techniques) or even the seismic amplification factor (micro-noise horizontal to vertical spectral ratio - HVSR) (Sarris et al., 2010).

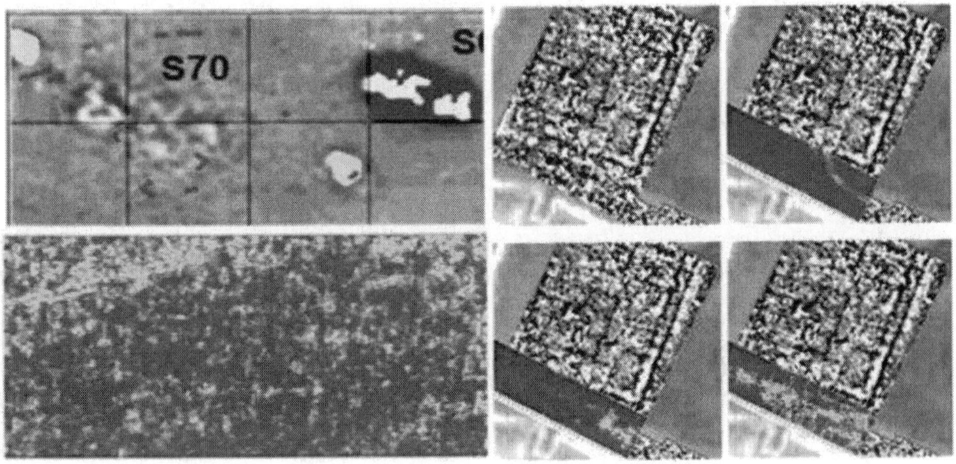

Figure 25. Left: Comparison between magnetic and GPR prospection above structural remains of the flat settlement at Szeghalom site in East Hungary. Even though the foundations of the daub constructions are registered clearly to the magnetic data (left top), the high conductivity of the soils has attenuated strongly the GPR electromagnetic sig- nals masking completely the particular area (left bottom) (Sarris 2012). Right: Comparison between magnetic and GPR prospection at the corner

of the Palaeochristian fortifications of Nikopolis, Epirus (Greece). The color maps represent the GPR horizontal slices of 0.1m width for depths of 0.5 (top right), 1 (bottom left) and 1.5m (bottom right) approxi- mately. The remains of a structural complex are obvious in the magnetic data. The GPR managed to register reflectors originating from various depths, such as a curving path at the top layers and a section of decumanus maximus at the lower bottom of the surveyed area. The latter was not clearly resolved in the magnetic data as the high surface con- centration of sherds created a uniform magnetic background masking of the area of interest.

Figure 26. Left: A 2-D view of the bedrock depth in the area of the harbour of Priniatikos Pyrgos resulting from the seismic refraction survey. The bluish colors indicate the deeper level of the bedrock and the dashed lines indicate the proposed location of the depression of the ancient harbour. Right: The soil resistance survey to the south of the prom- ontory of Priniatikos Pyrgos indicated a 5m wide high resistance linear anomaly that extends in a SW-NE direction and is probably related to one of the older branches of the Istron River running towards the east side of the promontory. (Sarris et al 2012)

Although current trends have emphasized the fast reconnaissance of the archaeological sites through multi-sensor, multi-electrode or multi-antenna systems, the manifold approach, which is the amalgamation of multiple geophysical techniques, as well as the fusion of the geophysical data with other types of remote sensing techniques, such as satellite imagery, LIDAR or laser scanning and orthophotos aiming towards a better and more holistic visualization of the area and a better reconstruction of the underground monuments will continue to be of crucial importance in the geophysical prospection of ar- chaeological context (Sarris 2012).

7. Low altitude systems for supporting archaeological investigations

Further to satellite and ground investigations, research has indicated the need for a low altitude airborne imaging systems in order to support archaeological research. This is due to the fact that such systems of low cost, with a stable platform for imaging sensors and have the ability to lift a payload equivalent to sensor equipment (Patterson & Brescia, 2008; Voer- hoeven, 2009; Kemper, 2012; Nebiker et al., 2008; Bento, 2008; Georgopoulos, 1982;

Hailey, 2005). In this study, several technologies were merged to create an innovative low altitude airborne system supporting remote sensing and photogrammetric applications, which includes the ability to conduct spectroscopy and aerial photography using a helium filled balloon. The complete low altitude airborne system is shown in Figure 27.

Figure 27. Right- ground control mechanism and aerial platform. Left-Low altitude airborne system including air balloon, spectro-radiometer, and researcher wearing ground control mechanism with harness (Themistocleous et al., 2012b)

A helium-filled balloon with a 3 m. diameter was used which was able to be raised to a height up to 200 m with a payload of up to 6kg. The Spectra Vista GER 1500 spectroradiometer was attached to the aerial platform and operated remotely. The balloon was raised to varying heights and spectroradiometric measurements were taken of the same target at different elevations. Concurrent to the spectroradiometric measurements, aerial photographs were taken using two digital cameras, one with infrared filter. The integration of the various techniques was used in order to detect subsurface archaeological remains by examining ground anomalies identified through spectral signatures. Previous campaigns in Cyprus found that field spectroscopy can support the detection of archaeological crop marks based on the retrieved spectral signatures over agricultural areas which are characterized as archeological areas (see Agapiou and Hadjimitsis 2011). Possible identification of subsurface archaeological remains is based on spectral signatures anomalies. Such anomalies are observed in crops when the vegetation is under stress due to subsurface relics. Therefore, spectral signatures anomalies are expected in the red and VNIR part of the spectrum.

The low altitude airborne imaging system was tested at the Agricultural Research Institute in Paphos, Cyprus, where a simulated archaeological test field was constructed. Spectroradiometric measurements and photographs in the visible and infrared range were taken over the target area. Preliminary results found that there were no significant differences in the spectral signatures in the visible range, while there was a significant difference among the spectral signatures in the NIR range as the balloon was moving up-wards (Figure 28). The study found that the spectral signature of the target can changed as a function of altitude, with higher reflectance indicated as the elevation increased.

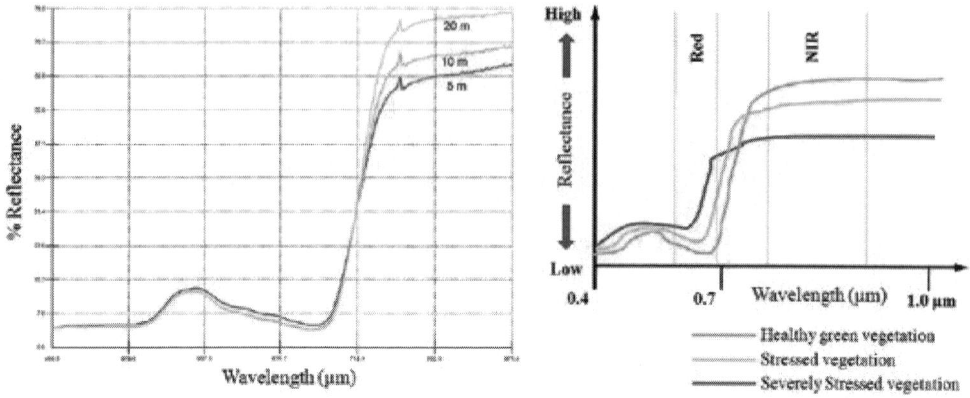

Figure 28. Right-Spectral signatures of vegetation at 5, 10 and 20 meters. Left-spectral differences between healthy and stressed vegetation (Themistocleous et al., 2012b)

8. Conclusions

Remote sensing can contribute in several ways to archaeological research. This chapter presents some results from different cases studies in Cyprus, Greece and Hungary using several techniques of remote sensing, including satellite images, archive aerial images, geo-physical surveys, 3D terrestrial laser scanners, ground spectroscopy, atmospheric pollution, WebGIS and GIS analysis for monitoring purposes.

The results have shown the potential use of satellite remote sensing and ground spectroscopy for the identification of buried archaeological remains through crop marks. Moreover, monitoring archaeological sites and risk assessment can be performed for several threats including urban expansion and air pollution. As demonstrated in this chapter, a dramatic land use change has taken place in several archaeological sites during the last decades. Such investigations are very important for studying archaeolandscapes since can provide valuable for information for areas that are nowadays vanished. Furthermore, the potential use of ground geophysical surveys for the detection of subsurface remains was also demonstrated through several applications in Greece and Hungary, was also demonstrated. Documentation, mapping. 3D modelling and WebGIS applications for archaeological sites and monuments are also demonstrated in this chapter.

References

1. Agapiou, A, Hadjimitsis, D. G, Alexakis, D, & Sarris, A. (2012a). Observatory validation of Neolithic tells ("Magoules") in the Thessalian plain, central Greece, using hy- perspectral spectro-radiometric data, *Journal of Archaeological Science*, doi. org/10.1016/ j.jas.2012.01.001., 39(5), 1499-1512.
2. Agapiou, A, Hadjimitsis, D. G, Alexakis, D, & Papadavid, G. (2012b). Examining the phenological cycle of barley (hordeum vulgare) using satellite and in situ spectrora-

diometer measurements for the detection of buried archaeological remains, *GIScience & Remote Sensing* 49 (6), 854-872.

3. Agapiou, A, Hadjimitsis, D. G, Sarris, A, Georgopoulos, A, & Alexakis, D. D. (2012c). Linear Spectral Unmixing for the detection of Neolithic Settlements in the Thessalian Plain, *Proceedings of the 32nd EARSeL Symposium*, Mykonos, Greece, May 2012.

4. Agapiou, A, Hadjimitsis, D. G, Papoutsa, C, Alexakis, D. D, & Papadavid, G. (2011). The importance of accounting for atmospheric effects in the application of NDVI and interpretation of satellite imagery supporting archaeological research: the case stud- ies of Palaepaphos and Nea Paphos sites in Cyprus. *Remote Sensing,* 3(12), 2605-2629,doi:10.3390/rs3122605

5. Agapiou, A, Hadjimitsis, G. D, Sarris, A, Themistocleous, K, & Papadavid, G. (2010a). Hyperspectral ground truth data for the detection of buried architectural re- mains, *Lecture Notes in Computer Science,* , 6436, 318-331.

6. Agapiou, A, Hadjimitsis, D. G, & Themistocleous, K. (2010b). Geometric documentation of historical churches in Cyprus using laser scanner, *Archaeolingua,* 978-9-63991-116-1*Proceedings of the 3rd International Euro-Mediterranean Conference (Eu- roMed),* Limassol, Cyprus, November 2010., 1-6.

7. Agapiou, A, Georgopoulos, A, Ioannides, C, & Ioannides, M. (2010c). A digital atlas for the Byzantine and Post Byzantine Churches of Troodos region (central Cyprus), *Proceedings of the CAA 2010 Conference,* Granada, Spain, April 2010.

8. Alexakis, D, Sarris, A, Astaras, T, & Albanakis, K. (2011). Integrated GIS, remote sensing and geomorphologic approaches for the reconstruction of the landscape hab- itation of Thessaly during the Neolithic period. *Journal of Archaeological Science,* 38(1), 89-100.

9. Alexakis, D, Sarris, A, Astaras, T, & Albanakis, K. (2009). Detection of Neolithic set- tlements in Thessaly (Greece) through multispectral and hyperspectral satellite im- agery. *Sensors,* doi:10.3390/s90201167., 9(2), 1167-1187.

10. Alexakis, D. & Sarris, A, (2010). Environmental and Human Risk Assessment of the Prehistoric and Historic Archaeological Sites of Western Crete (Greece) with the Use of GIS, Remote Sensing, Fuzzy Logic and Neural Networks, In Lecture Notes in Computer Science No. 6436: Digital Heritage (Third International Conference, Eu- roMed 2010, Lemessos, Cyprus, November 8-13, 2010 Proceedings) - Remote Sensing for Archaeology and Cultural Heritage Management and Monitoring, ed. by. Mari- nos Ioannides Dieter Fellner, Andreas Georgopoulos & Diofantos G. Hadjimitsis, Springer, 2010, pp. 332-342

11. Altaweel, M. (2005). The use of ASTER satellite imagery in archaeological contexts. *Archaeological Prospection,* , 12, 151-166.

12. Aqdus, S. A, Drummond, J, & Hanson, W. S. (2008). Discovering archaeological crop- marks: a hyperspectral approach. *Proceedings of The International Archives of the Photo- grammetry, Remote Sensing and Spatial Information Sciences, Vol. XXXVII. Part B5,* Beijing, China, July 2008.

13. Ayad, Y. (2005). Remote Sensing and GIS in modelling visual landscape change: a case study of the north-western arid coast of Egypt. *Landscape and Urban Planning*, 73, 4, , 307.

14. Barlindhaug, S, Holm-olsen, I. M, & Tømmervik, H. (2007). Monitoring archaeological sites in a changing landscape-using multitemporal satellite remote sensing as an 'early warning' method for detecting regrowth processes. *Archaeological Prospection*, doi:arp.307, 14, 231-244.

15. Bassani, C, Cavalli, R. M, Goffredo, R, Palombo, A, Pascucci, S, & Pignatti, S. (2009). Specific spectral bands for different land cover contexts to improve the efficiency of remote sensing archaeological prospection: the Arpi case study. *Journal of Cultural Heritage*, , 10, 41-48.

16. Beck, A. (2007). Archaeological site detection: the importance of contrast. *Proceedings of the Annual Conference of the Remote Sensing and Photogrammetry Society*, Newcastle University, Newcastle, United Kingdom, September, 2007.

17. Beck, A, Wilkinson, K, & Philip, G. (2007). Some techniques for improving the detection of archaeological features from satellite imagery. *Proceedings of the International Society for Optical Engineering, Remote sensing for environmental monitoring, GIS applica- tions, and geology VII*, Florence, Italy, September 2007,, 6749

18. Bento, M. D. F. (2008). Unmanned aerial vehicles: an overview. *Inside GNSS* (January/ February), , 54-61.

19. Bewley, R, Donoghue, D, Gaffney, V, Van Leusen, M, & Wise, A. (1999). *Archiving aerial photography and remote sensing data : a guide to good practice.* Archaeology Data Service, Oxbow, UK.

20. Capper, J. E. (1907). Photographs of Stonehenge as seen from a war balloon. *Archaeologia* , 60, 571.

21. Cavalli, R. M, & Colosi, F. Palombo, A; Pignatti, S. & Poscolieri, M. ((2007). Remote hyperspectral imagery as a support to archaeological prospection, *Journal of Cultural Heritage*, , 8, 272-283.

22. Carlon, C, Marcomini, A, Fozzati, L, Scanferla, P, Bertazzon, S, Bassa, S, Zanovello, F, Stefano, F, Chiarlo, R, & Penzo, F. (2002). ArcheoRisk: a decision support system on the environmental risk for archeological sites in the Venice lagoon. *Proceedings of the 1st Biennial Meeting of the iEMSs, Lugano, Switzerland.*

23. Comfort, A. (1997). Satellite remote sensing and archaeological survey on the Euphrates, *Aerial Archaeology Research Group News*, , 14, 39-46.

24. Crawford, O. G. S. (1923). Stonehenge from air: course and meaning of 'The Avenue', *Observer*, 13

25. Douglas, C. (2005). History and Status of Aerial and Satellite Remote Sensing and GIS in the Inventory and Evaluation of Cultural Sites, *Proceedings of the 8th Annual US/Icomos International Symposium Heritage Interpretation*, May 2005.

26. Fowler, M, & Curtis, H. (1995). Stonehenge from 230 kiliometers. *Aerial Archaeology Research Group News*, , 11, 8-16.

27. Fowler, M. J. F, & Fowler, M. Y. (2005). Detection of archaeological crop marks on declassified CORONA KH-4B intelligence satellite photography of Southern England. Archaeological Prospection, 12, 257-264.

28. Georgopoulos, A. (1982). Balloon and kite photography: An historical review. *International Archives of Photogrammetry*V 1, , 24, 196-206.

29. Grosse, G, Schirrmeister, L, Kunitsky, V. V, & Hubberten, H. W. (2005). The use of CORONA images in remote sensing of periglacial geomorphology: an illustration from the NE Siberian coast. *Permafrost and Periglacial Processes*, , 16(2), 163-172.

30. Glueck, N. (1965). *Deities and Dolphins: the story of the Nabataea.*, New York, Farrar, Straus and Gioroux.

31. Gourley, B, Lolos, Y, & Sarris, A. (2008). Application of integrated geophysical prospection techniques for mapping ancient Sikyon, Greece, *Proceedings of the 1st International Workshop on "Advances in Remote Sensing for Archaeology and Cultural Heritage Management", EARSEL,* Rome, Italy, September-October, 2008.

32. Hadjimitsis, D. G, Agapiou, A, Alexakis, D, & Sarris, A. (2011). Exploring natural and anthropogenic risk for cultural heritage in Cyprus using remote sensing and GIS, *In- ternational Journal of Digital Earth,* DOI:10.1080/17538947.2011.602119, 1-28.

33. Hadjimitsis, D. G, Agapiou, A, & Sarris, A. (2010). Risk assessment intended for cul- tural heritage sites and monuments in Cyprus using remote sensing and GIS, *Pro- ceedings of the 8th International Symposium on the Conservation of Monuments in the Mediterranean Basin,* University of Patras, Patras, Greece, May-June 2010.

34. Hadjimitsis, D. G, Retalis, A, & Clayton, C. R. I. (2002). The assessment of atmospheric pollution using satellite remote sensing technology in large cities in the vicinity of airports. *Water, Air & Soil Pollution: Focus, An International Journal of Environmental Pollution*, 2, 631-640.

35. Hadjimitsis, D. G, Themistocleous, K, Agapiou, A, & Clayton, C. R. I. (2009). Multi-temporal study of archaeological sites in Cyprus using atmospheric corrected satellite remotely sensed data. *International Journal of Architectural Computing,* 7(1), 121-138.

36. Hadjimitsis, D. G, Themistocleous, K, Ioannides, M, & Clayton, C. R. I. (2008). Integrating satellite remote sensing and spectro-radiometric measurements for monitoring archaeological site landscapes. *Proceedings of The 14th International Conference on Virtual Systems and Multimedia, VSMM*, Limassol, Cyprus, October 2008.

37. HadjimitsisD.G; Themistocleous, K.; Ioannides, M. & Clayton, C.R.I. ((2006). The reg- istration and monitoring of cultural heritage sites in the Cyprus landscape using GIS and satellite remote sensing. *Proceedings of the 37th CIPA International Workshop on e- Documentation and Standardization in Cultural Heritage, Symposium of CIPA, the ICO- MOS & ISPRS Committee on Documentation of Cultural Heritage*, Nicosia, Cyprus, October, 2006.

38. Hadjimitsis, D. G, & Themistocleous, K. (2008). The importance of considering atmospheric correction in the pre-processing of satellite remote sensing data intended for the management and detection of cultural sites: a case study of the Cyprus area.

Proceedings of the 14th International Conference on Virtual Systems and Multimedia (dedi- cated to culture heritage)- VSMM 2008, Limassol, Cyprus, October 2008.

39. Hailey, T. I. (2005). The powered parachute as an archaeological reconnaissance vehicle. *Archaeological Prospection*, 12, 69-78.
40. Johnson, J. K. (2006). *Remote Sensing in Archaeology*, The University of Alabama Press, Tuscaloosa, Alabama.
41. Kaufman, Y. J, Fraser, R. S, & Ferrare, R. A. (1990). Satellite measurements of large-scale air pollution: methods, *Journal of Geophysics Research*, 95, 9895-9909.
42. Kauth, R. J, & Thomas, G. S. (1976). The tasseled Cap- A Graphic Description of the Spectral-Temporal Development of Agricultural Crops as Seen by LANDSAT. *Proceedings of the Symposium on Machine Processing of Remotely Sensed Data*, Purdue Uni- versity of West Lafayette, Indiana, 4B, 44-51.
43. Kemper, G. (2012). New airborne sensors and platforms for solving specific tasks in remote sensing, *Proceedings of the International Archive of Photogrammetry, Remote Sens- ing and Spatial Information Science,* XXII ISPRS Congress, August-September 2012, Melbourne, Australia., XXXIX-B5
44. Keneddy, A. (1925). *Petra: Its history and monuments*. London, Country Life.
45. Keneddy, D. (2002). Aerial photography in the Middle East: the role of the military: past, present….and future?. In: *Aerial Archaeology: Developing Future Pactice*, Brewley, R.H. & Raczkowski, R. (eds), NATO Science book series, 1 Amsterdam, IOS Press.
46. Kostka, R. (2002). The world mountain Damavand: documentation and monitoring of human activities using remote sensing data. *ISPRS Journal of Photogrammetry and Remote Sensing*, 57(1-2), 5-12.
47. Laet, V, Paulissen, E, & Waelkens, M. (2007). Methods for the extraction of archaeological features from very high-resolution Ikonos-2 remote sensing imagery, Hisar (southwest Turkey), *Journal of Archaeological Science*, 34, 830e841.
48. Lanza, G.S., (2003). Flood hazard threat on cultural heritage in the town of Genoa (Italy), Journal of Cultural Heritage, vol. 4 (3), pp. 159-167.
49. Lasaponara, R, Masini, N, Rizzo, E, & Orefici, G. (2011). New discoveries in the Pira- mide Naranjada in Cahuachi (Peru) using satellite, Ground Probing Radar and mag- netic investigations. *Journal of Archaeological Science*, 38(9), 2031-2039.
50. Lasaponara, R, & Masini, N. (2007a). Detection of archaeological crop marks by using satellite QuickBird multispectral imagery. *Journal of Archaeological Science*, 34, 214-221.
51. Lasaponara, R, & Masini, N. (2007b). Improving satellite QuickBird-based identification of landscape archaeological features through Tasseled Cap Transformation and PCA. *Proceedings of the 21nd CIPA Symposium*, Athens, 2007.
52. Levin, N. (1999). *Fundamental of Remote Sensing*, Tel Aviv University, Israel.
53. Linford, N. (2006). The Application Of Geophysical Methods To Archaeological Prospection.. *Reports on Progress in Physics*, 69, 2205-2257.

54. Lock, G. (2003). *Using Computers in Archaeology, towards virtual pasts.* Routledge, Tay- lor and Francis Group, London and New York.

55. Masini, N, & Lasaponara, R. (2007). Investigating the spectral capability of Quickbird data to detect archaeological remains buried under vegetated and not vegetated areas, *Journal of Cultural Heritage*, 8, 53-60.

56. Mccauley, J. F, Schaber, G. G, Breed, C. S, Grolier, M. J, Haynes, C. V, Issawi, B, Ela- chi, C, & Blom, R. (1982). Subsurface valleys and geoarchaeology of the eastern Sa- hara revealed by Shuttle Radar. *Science*, 218, 1004-1020.

57. Monahan, E, & Sarris, A. (2011). Matters of Integration and Scale: New Efforts in Magnetometry Data Management at a Late Neolithic Settlement Site in Hungary, *Proceedings of the 16th International Congress "Cultural Heritage and New Technologies"*, Vienna, Austria, November 2011.

58. Nebiker, S, Annen, A, Scherrer, M, & Oesch, D. (2008). A Light-weight multispectral sensor for micro UAV opportunities for very high resolution airborne remote sens- ing. *The International Archive of Photogrammetry, Remote Sensing and Spatial Information Science*, 37, 1193-1200.

59. Negria, S, & Leucci, G. (2006). Geophysical investigation of the Temple of Apollo (Hi- erapolis, Turkey). *Journal of Archaeological Science*, 33(11), 1505-1513.

60. Nisantzi, A, Hadjimitsis, D. G, & Alexakis, D. (2011). Estimating the relationship be- tween aerosol optical thickness and PM10 using Lidar and meteorological data in Li- massol, Cyprus. *Proceedings of the SPIE Remote Sensing 2011*, 8182, Prague September, 2011.

61. Papadopoulos, N, Sarris, A, Yi, M-J, & Kim, J. H. (2009). Urban archaeological investigations using surface 3d ground penetrating and electrical resistivity tomography methods, *Exploration Geophysics*, 40, 56-68.

62. Papadopoulos, N, Tsourlos, P, Papazachos, C, Tsokas, G, Sarris, A, & Kim, J. H. (2011). An algorithm for fast 3-d inversion of surface electrical resistivity tomography data: application on imaging buried antiquities, *Geophysical Prospecting, EAGE*, 59, 557-575.

63. Parcak, S. H. (2009). *Satellite Remote Sensing for Archaeology*, Routledge Taylor and Francis Group, London and New York.

64. Patterson, M. C. L, & Brescia, A. (2008). Integrated sensor systems for UAS, *Proceedings of the 23rd Bristol International Unmanned Air Vehicle Systems (UAVS) Conference*, Bristol, United Kingdom, April 2008.

65. Raj, U, Poonacha, K. P, & Diga, S. (2005). Ruins of Hampi from high resolution remote sensing data- a case study. In: *Remote Sensing and Archaeology*, Tripathi (ed.), Sundeep Prakashan, New Delhi, 90-98.

66. Retalis, A, Hadjimitsis, D. G, Chrysoulakis, N, Michaelides, S, & Clayton, C. R. I. (2010). Comparison between visibility measurements obtained from satellites and ground, *Natural Hazards and Earth System Sciences Journal*, 10, 421-428.

67. Retalis, A. (1998). *Study of atmospheric pollution in Large Cities with the use of satellite observations: development of an atmospheric correction algorithm applied to polluted Urban areas*, PhD Thesis, Department of Applied Physics, University of Athens.

68. Retalis, A, Cartalis, C, & Athanasiou, E. (1999). Assessment of the distribution of aer- osols in the area of Athens with the use of Landsat TM. *International Journal of Remote Sensing*, 20, 939-945.

69. Riley, D. N. (1987). *Air photography and archaeology*. Duckworth, London.

70. Rowlands, A, & Sarris, A. (2006). Detection of exposed and subsurface archaeological remains using multi- sensor remote sensing. *Journal of Archaeological Science*, 34, 795-803.

71. Sabins, F. (1997). *Remote Sensing, Principles and Interpretation*. W.H. Freeman and Company, New York.

72. Sarris, A. (2008). Remote Sensing Approaches/Geophysical. In *Encyclopedia of Archaeology*, Rearsall, D. M., (ed.), Academic Press, New York, 3, 1912-1921.

73. Sarris, A. (2012). Multi+ or Manifold Geophysical Prospection? *Proceedings of Comput- er applications and Quantitative methods in Archaeology 2012*, University of Southamp- ton, Southampton, United Kingdom, March 2012.

74. Sarris, A, & Jones, R. (2000). Geophysical and Related Techniques Applied to Archaeological Survey in the Mediterranean: A Review", *Journal of Mediterranean Archaeology (JMA)*, 13, 3-75.

75. Sarris, A, & Papadopoulos, N. (2011). Geophysical Surveying in Urban Centers of Greece. *Proceedings of the 16th International Congress "Cultural Heritage and New Tech- nologies"*, Vienna, Austria, November 2011.

76. Sarris, A, Loupasakis, C, Soupios, P, Trigkas, V, & Vallianatos, F. (2010). Earthquake vulnerability and seismic risk assessment of urban areas in high seismic regions: application to Chania City, Crete Island, Greece, *Natural Hazards*, 54, 395-412.

77. Sarris, A, Papadopoulos, N, & Soupios, P. (2012). The Contribution of Geophysical Approaches to the Study of Priniatikos Pyrgos, *Proceedings of the Conference on Field-work and Research at Priniatikos Pyrgos and Environs 19122012*British School of Athens, Athens, Greece, June 2012.

78. Sarris, A, Papadopoulos, N, Theodoropoulos, S, Gourley, B, Shen, G, & Lolos, Y. Kal- paxis, Th. (2009). Revealing the Ancient City of Sikyon through the Application of In- tegrated Geophysical Approaches and 3D modelling, Proceedings of the 8th International Conference on Archaeological Prospection & 7th Colloque GEOFCAN, Memoire duSol, Espace des Hommes, Paris, France, September 2009.

79. Shahrukh, M, Soupios, P, Papadopoulos, N, & Sarris, A. (2012). Medium depth geophysical investigations at the Istron archaeological site, eastern Crete, Greece using 3D Refraction Seismic and Geoelectrical Tomography, (in press).

80. Skoulikides, N. T. (2000). *Corrosion and maintenance of construction materials of monu- ments*, Heraklion, (in greek).

81. Themistocleous, K, Nisantzi, A, Hadjimitsis, D. G, Retalis, A, Paronis, D, Michae- lides, S, Chrysoulakis, N, Agapiou, A, Giorgousis, G, & Perdikou, S. (2012a). Long term monitoring of air pollution in the vicinity of cultural heritage sites in Cyprus using remote sensing techniques. *International Journal of Heritage in the Digital Era*, 1(1), 145-168.

82. Themistocleous, K, Hadjimitsis, D. G, Georgopoulos, A, Agapiou, A, & Alexakis, D. D. (2012b). Development of a Low Altitude Airborne Imaging System for Supporting Remote Sensing and Photogrammetric Applications: 'The ICAROS Project' intended for archaeological applications in Cyprus, *Proceedings of the 4th International Euro-Mediterranean Conference (EuroMed 2012)*, Limassol, Cyprus October-November 2012 (in press).

83. Themistocleous, K. (2011). *Improving atmospheric correction methods for aerosol optical thickness retrieval supported by in-situ observations and GIS analysis.* Phd Thesis, Cyprus University of Technology, Department of Civil Engineering and Geomatics.

84. Tømmervik, H. (1995). Monitoring the effects of air pollution on terrestrial ecosystems in Varanger (Norway) and Nikel-Pechenga (Russia) using remote sensing. *Science of The Total Environment,* 160-16 , 753-767.

85. Tripathi, A. (2005a). Remote Sensing: an Introduction. In: *Remote Sensing and Archaeology*, Tripathi (ed.), Sundeep Prakashan, New Delhi, 1-10.

86. Tripathi, A. (2005b). Identification of Dvaraka through satellite remote sensing. In: *Remote Sensing and Archaeology*, Tripathi (ed.), Sundeep Prakashan, New Delhi , 78-86.

87. Urhus, M.P., Sullivan, P.A, & Mink, B. Ph., (2006), Identifying at-risk heritage resources with GIS: modelling the impact of recreational activities on the archaeological record, International Journal of Risk Assessment and Management, vol. 6 (4-6),330-343.

88. Vaughn, S, & Crawford, T. (2009). A predictive model of archaeological potential: An example from northwestern Belize. *Applied Geography*, 29(4), 542-555.

89. Ventera, M. L, Thompson, V. D, Reynolds, M. D, & Waggoner, J. C. (2006). Integrating shallow geophysical survey: archaeological investigations at Totogal in the Sierra de los Tuxtlas, Veracruz, Mexico. *Journal of Archaeological Science*, 33(6), 767-777.

90. Voerhoeven, G. J. J. (2009). Providing an archaeological bird's eye view-an overall picture of ground-based means to execute low-altitude aerial photography (LAAP) in Archaeology. *Archaeological Prospection*, , 16

Chapter-4

Integrated Remote Sensing and GIS Applications for Sustainable Watershed Management: A Case Study from Cyprus

1. Introduction

Due to the highly complex nature of both human and physical systems, the ability to com- prehend them and model future conditions using a watershed approach has taken a geo- graphic dimension. Satellite remote sensing and Geographic Information Systems (GIS) technology have played a critical role in all aspects of watershed management, from assess- ing watershed conditions through modeling impacts of human activities to visualizing im- pacts of alternative scenarios (Tim & Mallavaram, 2003).

The extreme weather phenomena and global warming noted in recent years has demonstrat- ed the necessity for effective flood risk management models. According to this paradigm, a considerable shift has been observed from structural defense against floods to a more com- prehensive approach, including appropriate land use, agricultural and forest practices (Alexakis et al., 2013a, 2013b; Barredo & Engelen, 2010; Lilesand & Kiefer, 2010; Michaelides et al., 2009). Land cover changes may be used to describe the dynamics of urban settlements and vegetation patterns as important indicators of urban ecological environments (Yinxin & Linlin, 2010). Satellite remote sensing provides an excellent source of data from which up- dated land use / land cover (LULC) changes can be extracted and analysed in an efficient way. In addition, effective monitoring and simulating of the urban sprawl phenomenon and its effects on land-use patterns and hydrological processes within the spatial limits of a wa- tershed are essential for effective land-use and water resource planning and management (Hongga et al., 2010; Hadjimitsis et al., 2004a, 2010a, 2010b). Several techniques have been reported in order to improve classification results in terms of land use discrimination and accuracy of resulting classes in the processing of remotely sensed data (Agapiou et al., 2011). As a result of Very High Resolution (VHR) imagery, real world objects that were previously represented by very few pixels, are now represented by many pixels. Thus, techniques that take into account the spatial properties of an image region need to be developed and ap- plied. One such technique is texture analysis (Zhang & Zhu, 2011). Moreover, during the last years, spatial metrics have been largely used in landscape studies. According to Haralick et al. (1973), landscape metrics capture the inherent spatial structure of the environment and are used to enhance interpretation of spatial pattern of the landscape.

Several techniques have been reported to improve classification results in terms of land use discrimination and accuracy of resulting classes (Eiumnoh & Shrestha, 2000). However, the multispectral images acquired from different satellite sensors suffer from serious problems and errors, such as radiometric distortions, areas with low illumination, physical changes of the environment, etc. Recent studies have found that the accuracy of classification of remote sensing imagery does not increase by improving the applied algorithms, since classification mainly depends upon the physical and chemical parameters of the objects on the ground (Rongqun & Daolin, 2011).

Soil erosion is considered to be a major environmental problem, as it seriously threatens natural resources, agriculture and the environment in a catchment area. Spatial and quantitative information of soil erosion contributes significantly to the soil conservation management, erosion control and general catchment area management (Prasannakumar et al., 2011). In recent years, there has been a growing awareness of the importance of problems directly related to erosion in the broader Mediterranean region. The widespread occurrence and importance of accelerated erosion in the Mediterranean region has driven to the development of models at scales ranging from individual farm fields to vast catchment areas and different types of administrative areas (Bou Kheir et al., 2008). In some parts of the Mediterranean region, erosion has reached a stage of irreversibility, while in some places there is no more soil left (Kouli et al., 2009). Although soil erosion is characterized as a natural phenomenon, human activities such as agriculture can accelerate it further (Karydas et al., 2009).

Recently, space-born microwave active remote sensing, especially Synthetic Aperture Radar (SAR) with its all-weather capability, can provide useful spatially distributed flood information that may be integrated with flood predictive models in the construction of an effective watershed management. Radar imagery is useful for the identification, mapping and measurement of streams, lakes and inundated areas. Most surface water features are detectable on radar imagery due to the contrast between the smooth water surface and the rough land surface (Lewis, 1998). The amount of moisture stored in the upper soil layer changes the dielectric constant of the material and thus affects the SAR return. Because the dielectric constant of water is at least 10 times bigger than that of the dry soil, the presence of water in the top few centimeters of bare soil can easily be detected through the use of SAR imagery (Lil- lesand & Kiefer, 2000). In addition, the differences in the values between the dielectric constant of water and of dry soil at the microwave part of the spectrum plays a major role in the soil moisture estimation through the use of microwaves.

The main aim of this chapter is to integrate all the individual remote sensing methodologies related to watershed monitoring and management in a holistic approach. Specifically, different approaches such as development of erosion models, use of radar imagery for the detection of areas prone to inundation phenomena, construction of Land Use /Land Cover (LULC) maps, optimization of classification methodologies and calculation of landscape metrics for the recording of urban sprawl will be presented thoroughly and will highlight the contribution of satellite remote sensing to the sustainable management of a catchment area.

2. Study area

Located in the central part of the island of Cyprus, the Yialias basin is about 110 km^2 in size (Fig. 1). This study area is situated between longitudes 33°11′24.28′′ and 33°26′31.52′′ and latitudes 34°54′36.74′′ and 35°2′52.16′′. Cyprus is located in the Northeastern corner of the Mediterranean Sea and, therefore, has a typical eastern Mediterranean climate: the combined temperature–rainfall regime is characterized by cool-to-mild wet winters and warm- to-hot dry summers (see Michaelides et al., 2009).

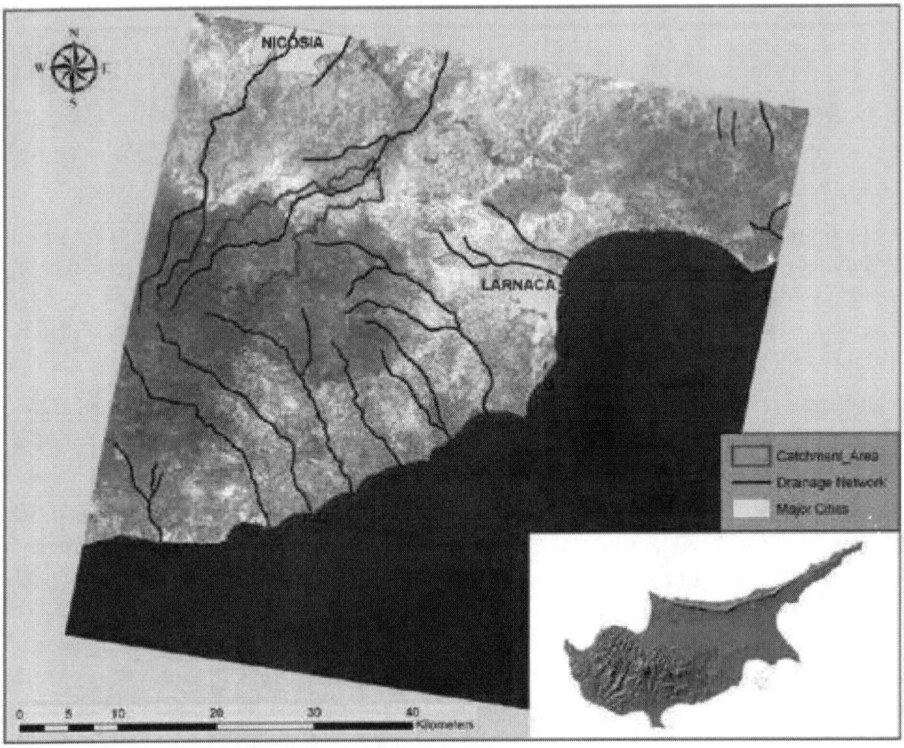

Figure 1. The study area

3. Development of methodology for the optimization of classification accuracy of Landsat TM/ETM+ imagery in a catchment area in Cyprus

3.1 Introduction

An important tool for the detection and quantification of land cover changes across catchment areas is the classification of multispectral satellite imagery, as such results are very im- portant for hydrological analysis and flood scenarios.

This study aimed at testing different material samples in the Yialias region (central Cyprus) in order to examine: a) their spectral behavior under different precipitation rates and b) to introduce an alternative methodology to optimize the classification results derived from single satellite imagery with the combined use of satellite, spectroradiometric and precipitation data.

3.2 Data and methodology

3.2.1. Ground sample

According to preliminary classification results (Alexakis et al., 2011), spectral mixing be- tween urban areas and specific geological formations was observed. Thus, samples of re- golith and construction material were collected and tested for their spectral response under different conditions of humidity with the use of spectroradiometer in the premises of the Remote Sensing and Geomatics Laboratory of Cyprus University of Technology (Alexakis et al., 2012).

3.2.2. Satellite and precipitation data

For the purposes of the study, specific tools and data were incorporated:

- ❖ Four Landsat TM/ ETM+ multispectral images of medium resolution (30x30 m^2 pixel size).
- ❖ Precipitation data obtained from the Meteorological Service of Cyprus (Pera Chorio Meteorological Station : Lon - 35° 01', Latitude - 33° 23'). All of these data were compared with the satellite imagery data. Selected satellite imagery was retrieved a day after the recording of substantial scaling amount of precipitation from the Pera-Chorio Metereological Station.
- ❖ Data derived from spectroradiometric field campaigns. For this reason a GER 1500 spec- troradiometer was used. This instrument can record electromagnetic radiation between 350 nm up to 1050 nm (Fig. 2).

In order to investigate the different spectral response of each sample under different moisture conditions, all samples were immersed in water in a step-by-step process and measured for the rate of their humidity with a soil moisture meter. The specific hand-held instrument used in this study was able to measure moisture values from 0 to 50% within an accuracy of 0.1%. The final under investigation regolith samples were divided in four different catego-ries, according to their level of humidity: 0% (dry sample); 25%; 50%; > 50%. With regard to tile and roof specimens, the results were divided into "dry" or "humid" categories due to the difficulty to measure the scaling levels of humidity in those kinds of materials.

Figure 2. Collection of soil data (left). Spectroradiometric measurements of material samples at the premises of the Remote Sensing and Geomatics Laboratory of CUT (right)

Based on the results of the scatter-plots, it was found that in the case of dry samples there is a strong spectral confusion between the chalk A response and the urban fabric (roof and tile) materials. The "moisture" scatter plot (humidity > 50%) highlights the different spectral re- sponse between artificial materials (roof and tile) and natural materials (chalk A, B, C). In this plot, the spectral difference between different samples is increased and two major clus- ters are created with complete contrary spectral response (increase of chalk A spectral re- sponse and substantial decrease of tile and house roof -constructed from clay and cement consecutively- spectral response, see Fig.3).

The results highlighted the different spectral response of materials under different humidity levels. Specifically, reflectance values of chalk samples (samples A and C) tend to be sepa- rated from those of urban samples (tile and roof) as humidity increases.

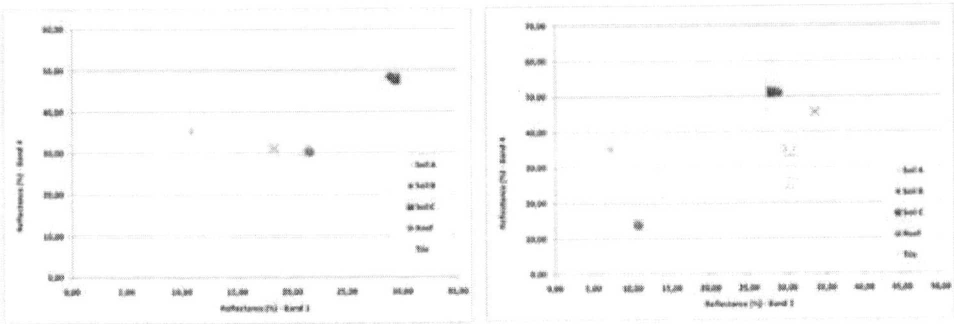

Figure 3. Scatter-plots of the different targets examined in this study for Band 1 - Band 4 (humidity 0%) (left) and Band 1 - Band 4 of Landsat (humidity > 50%) (right)

3.2.3 Satellite imagery data

After the application of all necessary pre-processing steps (radiometric, atmospheric and geometric corrections,) spectral signature profiles were extracted for all of the different materials during the acquisition dates of each satellite imagery (Fig. 4).

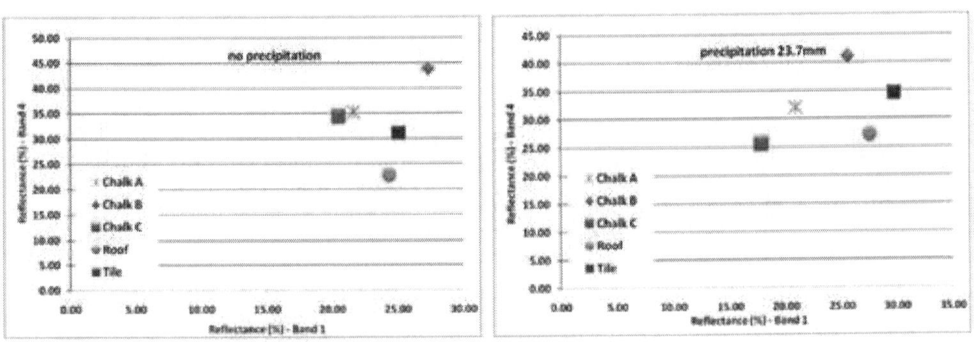

Figure 4. Scatter-plots of the different targets examined in this study for Band 1 - Band 4 (left) and Band 3 - Band 4 of Landsat (right)

The results of the scatter plots denoted the scaling optimization of spectral separability of satellite imagery data, from 0 to 23.7 mm of precipitation. Specifically, concerning the 0 mm precipitation case, a spectral confusion was indicated between the "urban" targets (roof and tile) and chalk A and C targets. This conflict was outreached gradually as the precipitation level increased. The samples started to have different spectral behaviour, with the chalk samples (except chalk B) standing gradually away from the "urban" sam- ples cluster in the scatter-plot. It is important to mention the quite different spectral re- sponse of chalk C sample in satellite images compared to its response in the laboratory specimens. This problem occurred due to the medium spatial resolution of Landsat im- ages (30x30 m^2 pixel size) which increases the likelihood of the common mixing pixel phenomenon.

3.3. Results and verification

The results from the laboratory and satellite imagery analysis methods highlighted the different spectral response of materials to different levels of humidity. For the direct comparison of the classification accuracy between images, where different levels of precipitation have been recorded, two Landsat TM/ETM+ images acquired on 2 June 2005 (0 mm precipitation – "dry") and 23 July 2009 (23.7 mm precipitation – "rainy") were clas- sified and compared (Fig. 5). Both unsupervised (ISODATA) and supervised classifica- tion algorithms (Maximum Likelihood - ML) were used. Initially, the ISODATA classification technique was applied to both images with 95% convergence threshold. The following 5 classes were used for both the supervised and unsupervised algorithms: 1) urban Fabric, 2) marl - chalk formations, 3) vegetation, 4) bare soil and 5) forest.

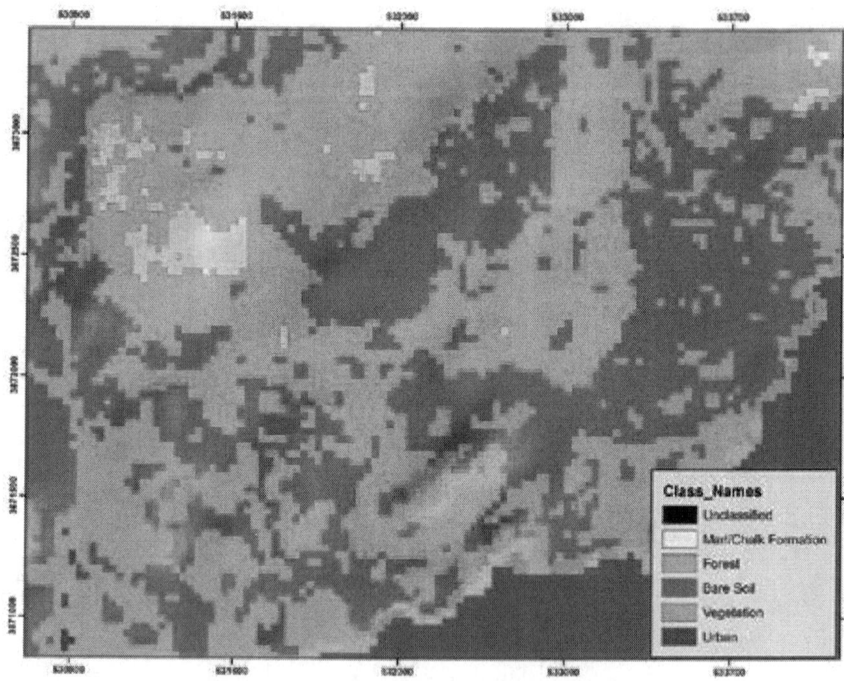

Figure 5. Detail of the "rainy" satellite image after the application of supervised classification algorithm

On the one hand, the results of the unsupervised algorithm performance for both dry and humid acquisition days could be described as poor and were not considered for further evaluation (Kappa coefficient of classification accuracy - (Kc) < 60%). On the other hand, the application of supervised algorithm to "rainy" image provided better accuracy results (Kc = 0.75). The product of "dry" image was substantially better than that of unsupervised case but with insufficient accuracy to be considered as credential.

3.4 Conclusions

The results noted the importance of imagery acquisition date for optimization of classification results. Specifically, the overall accuracy of classification product was substantially increased (more than 30% for supervised classification), especially for urban and marl/ chalk areas, during days where high precipitation measurements were recorded in the broader study area. The results were established either by laboratory or satellite imagery analysis.

4. Assessing soil erosion rate in a catchment area in Cyprus using remote sensing and GIS techniques

4.1. Introduction

The objective of this work was to develop and evaluate two different erosion models in the catchment area of Yialias in Cyprus. The first was an empirical multi-parametric model which is mainly based on expert's knowledge (Analytical Hierarchical Process - AHP) and the second (Revised Universal Soil Loss Equation - RUSLE) was the model which is consid- ered to be a contemporary simple and widely used approach of soil loss assessment.

4.2 Methodology

4.2.1. RUSLE methodology

The RUSLE equation incorporates five different factors concerning rainfall (R), soil erodibili- ty (K), slope length and steepness (L and S. respectively), support practice (P) and cover management (C):

$$A=R\ \ K\ \ L\ \ S\ \ P\ \ C \tag{1}$$

AHP allows interdependences between decision factors to be taken into account and uses expert opinions as inputs for evaluating decision factors. The final weight of significance for each factor can be defined by using the eigen-vectors of a square reciprocal matrix of pair- wise comparisons between the different factors. Moreover, a specific grade is assigned to all the different pairs from 1/9, when the factor is "not important at all", to 9, when the factor is "extremely important".

4.2.1.1. Rainfall (R) factor

The rainfall factor R is a measure of the erosive force of a specific rainfall value. For the cal- culation of the R factor with the use of the Modified Fournier Index (MFI), the following two different approaches suggested by Ferro et al. (1991) and Renard & Freimund (1994) for the areas of Sicily and Morocco were used respectively :

$$R_1 = 0.612\ MFI^{1.56} \tag{2}$$

$$R = 0.264\ MFI^{1.50} \tag{3}$$

According to Kouli et al. (2009), MFI is well correlated with the rainfall erosivity. The specific index is considered as an effective estimator of R because it takes into account the rainfall seasonal distribution. Therefore the MFI was applied to take into account the monthly rain- fall distribution during each year for a period of 20 years, as follows:

$$F_f = \sum_{j=1}^{N} \frac{Fa_j}{N} = \frac{1}{N} \sum_{j=1}^{N} \sum_{i=1}^{12} \frac{P_{ij}}{P_i} \tag{4}$$

where, F_f is the MFI index, p_{ij} is the rainfall depth in month / (mm) of the year j and P is the rainfall total for the same year. After the calculation of R, a continuous surface was pro- duced using the ordinary Kriging method based on Gaussian function, which was found to be the most effective for the production of the final iso-erosivity map. The mean values of R range from 267 MJ mm ha year^{-1} in the most flat areas in Yialias watershed to 694 MJ mm ha year^{-1} in the mountainous and generally steep areas.

4.2.1.2. Soil erodibility (K)

The soil erodibility factor (K) refers to the average long-term soil and soil profile response to the erosive power associated with rainfall and runoff. It is also considered to represent the rate of soil loss per unit of rainfall erosion index for a specific soil.

A digital soil map of the study area was used and the main soil formations were categorized in three different major classes: coarse sandy loam, sandy loam and silty clay. According to Prasannakumar et al. (2011) the estimated K values for the textural groups vary from 0.07 t ha h ha^{-1} MJ^{-1} mm^{-1} for coarse sandy loam, 0.13 t ha h ha^{-1} MJ^{-1} mm^{-1} for sandy loam and 0.26 t ha h ha^{-1} MJ^{-1} mm^{-1} for silty clay.

4.2.1.3. Topographic factor (LS)

The topographic factor is related to the slope steepness factor (S) and slope length factor (L) and is considered to be a crucial factor for the quantification of erosion due to surface run–off.

The combined topograpfic factor was calculated by means of ArcGIS spatial analyst and Hy- drotools extension tools. In this study, the equation derived from Moore & Burch (1986) has been adoped:

$$LS = \left(\frac{\text{Flow Accumulation} \cdot \text{CellSize}}{22.13}\right)^{0.4} \cdot \left(\frac{\sin(\text{Slope})}{0.0896}\right)^{1.3} \tag{5}$$

4.2.1.4. Practice factor (P)

The practice factor (P) is defined as the ratio of soil loss after a specific support practice to the cor- responding soil loss after up and down cultivation. In order to delineate areas with terracing practices, the two GeoEye-1 satellite images were used and the delineation was accomplished in GIS environment with extensive monitoring of the study area. Areas with no support practice were assigned with a P factor equal to 1. However, the terrace areas which are considered to be less prone to erosion were assigned a 0.55 value, according to expert's opinion.

4.2.1.5. Cover management factor (C)

According to Prasannakumar et al. (2011), the C factor represents the effect of soil-disturbing activities, plants, crop sequence and productivity level, soil cover and subsurface biomass on soil erosion.

The NDVI (Normalised Difference Vegetation Index) extracted from the study area (applied to GeoEye-1 image) has values that range from -0.65 to 0.99. The NDVI is used along with the Equation 6 in order to calculate the *C* factor values of the study area in GIS environment.

$$C=\exp -a\frac{NDVI}{(b-NDVI)} \tag{6}$$

where, a and b are non-dimensional parameters that determine the shape of the curve relat- ing to NDVI and C factor.

According to the final results, C factor values ranged from 0 to 2.7.

4.2.1.6. Application of RUSLE methodology for soil loss estimation

The annual soil loss was calculated in a GIS environment (Fig. 6), according to Eq. 1. According to the final results, the estimated soil loss ranges from 0 to 6394 t ha^{-1} yr^{-1} with a mean value of 20.95 t ha^{-1} yr^{-1}. The maximum value of 6394 t ha^{-1} yr^{-1} cannot be considered as appreciable due to the fact that only one pixel in a total of 1199 was attributed with this val- ue. However, the mean value of 20.95 t ha^{-1} yr^{-1} is representative of the current soil loss re- gime of the basin.

4.3. AHP methodology

In the AHP methodology, interdependencies and feedback between the factors were consid- ered. The factors used in this methodology were: rainfall (R), soil erodibility (K), slope length and steepness (LS), cover management (C), support practice (P) and stream proximi- ty. Six out of seven factors had already been analyzed in the RUSLE methodology. The addi- tional agent to be analyzed was the proximity to rivers and streams.

4.3.1. Proximity to rivers and streams

According to Nekhay et al. (2009), an area of 50 m around rivers and streams was considered to be prone to flooding and, consequently, to the detachment of particles of soil by floodwaters. Thus, initially with the use of ArcGIS 10 Hydrotools module, the drainage net- work of the basin was automatically extracted from the hydrological corrected DEM (Digital

Elevation Model). Next, a buffer zone of 50m was constructed around each drainage net- work segment.

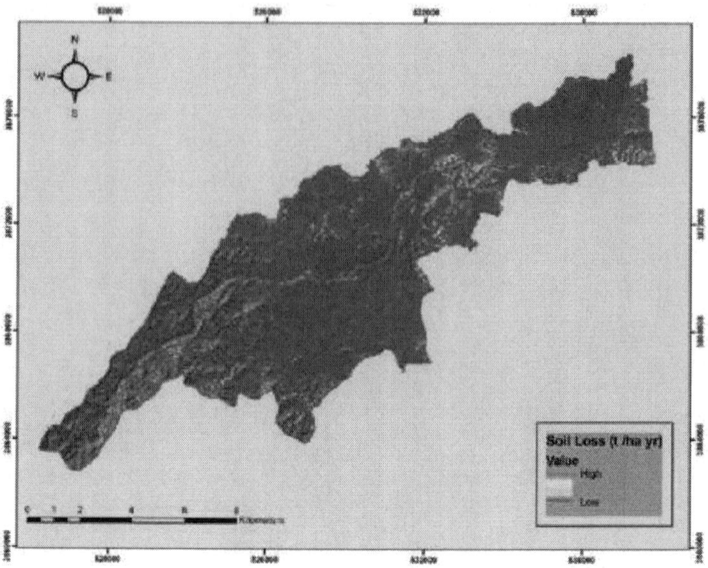

Figure 6. Map of the spatial distribution of soil loss after the application of RUSLE methodology in Yialias catchment area

According to AHP methodology, a pair-wise comparison of the contribution of each factor was established. Specifically, answers of several experts were collected on the reciprocal ma- trix, and the appropriate eigenvector solution method is then employed to calculate the fac- tor weights.

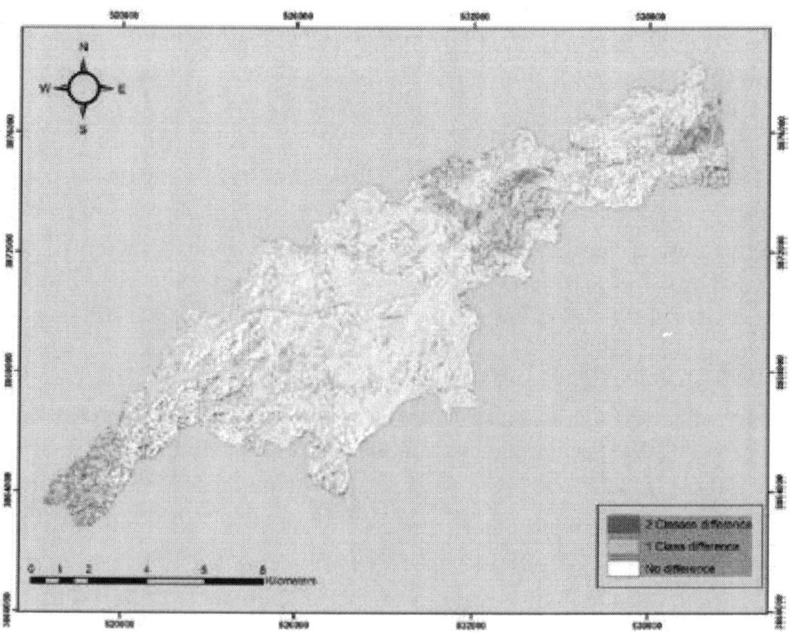

Figure 7. Final erosion risk map constructed with AHP method

The final soil erosion risk map (Fig. 7) was constructed by summing up (through Boolean operators) the product of each category (that had already been rated accordingly for its subcategories) with the corresponding weight of significance according to the following equation:

$$LS = F1 \quad 0.025 + F2 \quad 0.09 + F3 \quad 0.146 + F4 \quad 0.059 + F5 \quad 0.38 + F6 \quad 0.3 \qquad (7)$$

Where F1, F2,..., FN are the different factors incorporated in the model.

The final erosion risk assessment map was reclassified to three soil erosion severity classes separated as low (pixel value 1), moderate (pixel value 2) and high risk (pixel value 3). The results denoted that 77.5% of the study area was classified as low potential erosion risk, 17.5% as moderate potential risk and only a 5% as high risk.

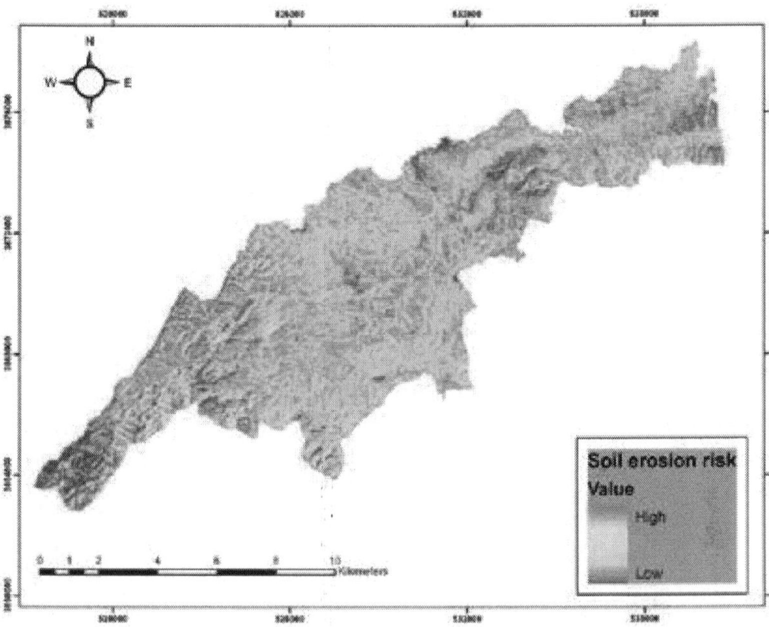

Figure 8. Image indicating the soil erosion severity class differences between AHP and RUSLE method

4.3.2. Evaluation of AHP and RUSLE

In the same way that the AHP risk assessment map was reclassified, the estimated soil loss percentage map was separated in 3 different classes according to experts opinion (1st class 0-20 t ha^{-1} yr^{-1} for pixel value 1, 2nd class 20-100 t ha^{-1} yr^{-1} for pixel value 2, 3rd class 100-6391 t ha^{-1} yr^{-1} for pixel value 3). The two grid images were subtracted in GIS environment. A close look at the extracted grid image, it is obvious that there is a considerable similarity between the two methodologies (Fig. 8).

4.4 Conclusions

This research demonstrated the potential for the integration of RS, GIS and precipitation da-ta to model soil erosion. The current research found that both RUSLE and AHP methodologies can be efficiently applied at a basin scale with quite modest data requirements in a Mediterranean environment such as Cyprus, providing the end users with reliable quantitative and spatial information concerning soil loss and erosion risk in general.

5. Flood mapping of Yialias river catchment area in Cyprus using ALOS PALSAR radar images

5.1 Introduction

ALOS (Advanced Land Observing Satellite) PALSAR data can be used to detect the water sur- face due to the L-band wave length. All SAR instruments share the advantages of day-night operability (as active sensors), cloud penetration, and the ability to calibrate without perform- ing atmospheric corrections. The longer L-band (~23.5 cm) SAR wavelength, and, to a certain extent, the C-band (~5.5 cm), have the ability to penetrate vegetation canopies to various de- grees depending on vegetation density and height, dielectric constant (primarily a function of water content), and SAR incidence angle. Variations in backscattering allow discrimination among non-vegetated areas (very low to low returns), herbaceous vegetation (low to moder- ate returns), and forest (moderate to high returns), and to some degree among different forest structures and regrowth stages. Where water is present beneath a forest canopy, enhanced re- turns caused by specular "double bounce" scattering between water surface and tree trunks makes it possible to distinguish between flooded and non-flooded forest.

5.2 Data and methodology

5.2.1. Data and methodology

The purpose of this study is to explore the potential of ALOS-PALSAR imagery for observing flood inundation phenomena in the Yialias catchment area in Cyprus. Two PALSAR images (polarity: HH, pixel size 50 m) covering the study area before and after an extreme precipita- tion incident in 2009 were used (Table 1). A LULC map was also constructed with the use of high resolution images such as GeoEye -1 covering the study area. To analyze Radar backscatter behavior for different land cover types, several regions of interest were selected based on the land cover classes. A number of land cover classes were found to be sensitive to flooding, whereas in some other classes backscatter signatures remained almost unchanged.

5.2.2. Data

For the purposes of the study, the following satellite and digital spatial data were incorporated:

- 2 ALOS PALSAR images.
- 2 GeoEye -1 images
- A Digital Elevation Model (DEM) of 25m pixel size provided by the Department of Land and Surveys of Cyprus, created with the use of orthorectified stereopairs of aerial photos covering the study area.

The ALOS images were acquired on 30 November 2009 and 6 December 2009 (Fig. 9a). PAL- SAR is a fully polarimetric instrument, operating at L-Band with 1270 MHz

(23.6 cm) centre frequency and 28 MHz, alternatively 14 MHz, bandwidth. The antenna consists of 80 transmit /receive (T/R) modules on four panel segments, with a total size of 3.1 by 8.9m (Table 1). The two ALOS images were acquired after thorough indexing of Cyprus Meteorological Service archives of precipitation data. Specifically, the research team searched the precipitation archives of all the meteorological and climatological gauge stations within the study area (Analiontas, Pera Chorio, Lythrodontas, Mantra tou Kampiou, Kionia, Mathiatis), as they are indicated and spatially distributed in Fig. 10. Due to the lack of ALOS imagery data acquired during recorded flood inundation events, the research team tried to acquire images before and after extreme precipitation events in order to examine the potential of the imagery to detect soil moisture and flood inundation trends. Thus, the image for 30 November 2009 corresponded to a day where no precipitation had been recorded, while the image for 6 December 2009 corresponded to a day when a mean value of 25mm of precipitation had been recorded in the rain gauge stations within the study area.

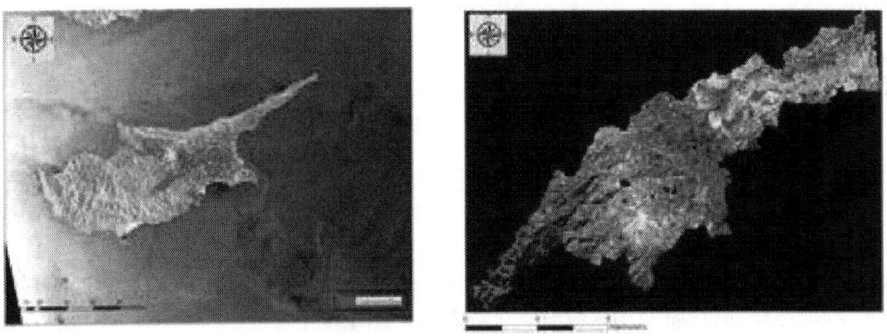

(a) (b)

Figure 9. (a) ALOS PALSAR image (30 November 2009) and the study area. (b) Mosaic of the two GeoEye -1 images of the study area (RGB - 321)

The GeoEye-1 images were used for the land use monitoring of the upstream and downstream of the basin; the two images refer to 12 March 2011 and 11 December 2011, respectively. GeoEye-1 is a multispectral sensor with four spectral bands. Its spectral range is: 450-510 nm (blue), 510-580 nm (green), 655-690 nm (red) and 780-920 nm (near infrared), while its spatial resolution is approximately 1.65 m.

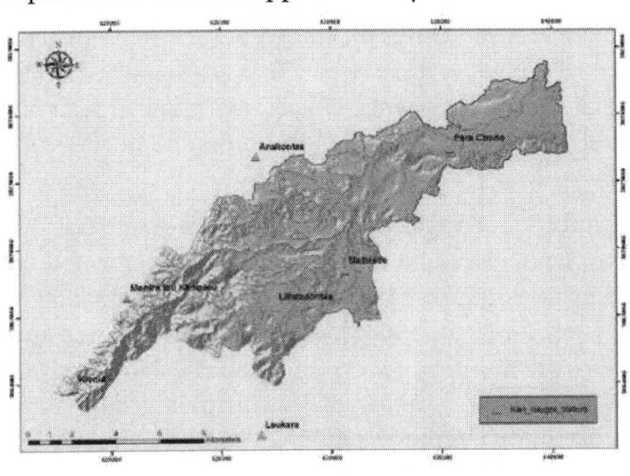

Figure 10. Rain gauge stations within the study area or in close vicinity with it and drainage network

Table 1. Technical specifications of ALOS PALSAR images

	ScanSar (WB1)
Resolution	50m
Swath Width	35 km
Polarization	HH
Off Nadir –Angle (deg)	18.0-43.3
Incidence Angle (deg)	20.1-36.5
Processing level	4.2
Data Rate (Mbps)	120
Bit quantization (bits)	5
Projection	UTM Zone 36 North

5.3. Pre-processing techniques

Initially, geometric corrections were carried out to ALOS PALSAR images using standard techniques with ground control points and a first order polynomial fit so as the two images to be co-registered. For this purpose, topographical maps were used to track the position of ground control points in conjunction with the digital shoreline of Cyprus extracted from the provided DEM. There are ascending and descending observation modes of PALSAR images and differences in backscattering values, therefore, the image calibration is an essential task. Different factors influence backscatter strength signal including satellite ground track, incidence angle, radar polarization, surface roughness and the surface's dielectric properties (Yingxin & Linlin, 2010). Different objects having the same digital number which may correspond to different backscatter values. Thus, the ALOS scenes were subsequently converted from amplitude data format to normalized radar cross section (σ°) according to Equation 8:

$$\sigma^{\circ} = 10 \log_{10} DN^2 + CF, \quad (8)$$

where, DN is Digital Number and CF is a calibration factor (CF = - 83.0 dB).

In SAR image, the speckle noise is one of obstacles to overcome in data processing, so it is necessary to take effective steps to filter the image. Several filter algorithms were tried; the Lee filter was applied to reduce speckle noise. This filter is based on the minimum mean square root (MMSE) and geometric aspects. This is a statistical filter designed to eliminate noise, while still maintaining the quality of pixel points and borders of the image (Hongga et al., 2010).

Atmospheric and geometric corrections were carried out on the GeoEye-1 images. Atmospheric correction is considered to be one of the most complicated techniques since the distributions and intensities of these effects are often inadequately known. Despite the variety of techniques used to estimate the atmospheric effect, the atmospheric correction remains a difficult task in the pre-processing of image data. As it is shown by several studies (Hadjimitsis et al. 2004b, 2010a, 2010b; Agapiou et al., 2011), the darkest pixel (DP) atmospheric correction methodology can easily be applied either by using dark targets located in the image or by conducting *in situ* measurements.

After the application of atmospheric and geometric corrections to GeoEye-1 images, the research team proceeded in the construction of an overall image mosaic by integrating the two individual images covering the up- and down-stream of the watershed basin (Fig. 9b). For this purpose, a histogram matching technique was applied to the common covered area of the two images in order to secure the radiometric correctness of the final extracted mosaic. Finally, the research team removed the cloud cover from the mosaic im- age in GIS environment.

5.4. GeoEye-1 Imagery classification and ALOS PALSAR texture analysis

5.4.1. GeoEye-1 Imagery classification technique

After the application of preprocessing techniques to GeoEye-1 images and the development of an image mosaic, the Maximum Likelihood (ML) algorithm was applied to create a de- tailed LULC map of the study area. For this reason 7 major classes were defined (Bare rock, Forest, Marl, Soil, Trees, Urban Fabric, Agricultural Areas) (Fig. 11). The statistics of the land use regime of the study area are shown in Table 2. From these statistics, it is clearly seen that the main part of the catchment area is covered by soil and olive trees.

Table 2. Statistics of the LULC thematic map

	Classes	Area (km²)
Bare Rock		2.52
Forest		3.99
Marl		0.33
Soil		43.86
Trees (mainly olive trees)		47.99
Urban Fabric		8.21
Agricultural Areas		2.99

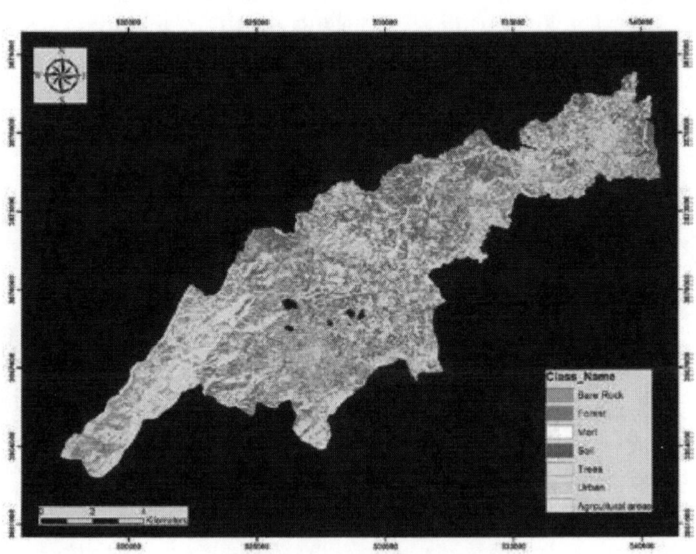

Figure 11. LULC map of the study area after the application of ML classification algorithm to GeoEye-1 mosaic

5.4.2. ALOS PALSAR texture analysis

According to Zhang & Zhu et al. (2011) texture is defined as the spatial variation in gray value and is independent of color or luminance. Texture measures smoothness, coarse- ness and regularity of a region in an image. For the description of texture histograms, gray level co-occurrence matrix (GLCM), local statistics and characteristics of the frequen- cy spectrum are used. The GCLM mainly operates by calculating a matrix that is based on quantifying the difference between the grey levels of neighboring pixels in an image window. The main aim of this matrix is the quantification of the spatial pixel structure within this window. It was initially suggested as a mechanism for extracting texture measures (Haralick et al., 1973).

In the specific study, through the use of ENVI 4.7 software, 7 different statistical indica- tors of texture such as contrast, angular second moment, homogeneity, entropy, dissimi- larity, mean and variance were applied for carrying out the statistical texture analysis of all the typical ground objects. From those textural indicators, multiple RGB composites were constructed to improve the visual monitoring and interpretation of moisture affect- ed areas.

5.5. Results and discussions

As it is clearly seen in Figure 12, in the downward of the catchment area (northeastern part) certain patches were inundated with water. Those patches are clearly observed with the low backscattering values and their corresponding dark pixels. However, in most of the cases the backscattering values were increased mainly because of volume scattering due to the moisture effect in the vegetation and plant cover. Concerning the southwestern part of the watershed where the most forested areas are established, due to the corresponding increase of the moisture after the extreme precipitation event, the backscatter values were generally increased due to the effect of double reflection by wa- ter (moisture) and tree trunks. Thus, generally the SAR backscattering intensity in forest areas changes to be higher in cases of inundation events. In addition, in certain areas of the southwestern part of the catchment area where there are more bare rock and soil patterns, the backscattering values were decreased due to the corresponding moisture effect.

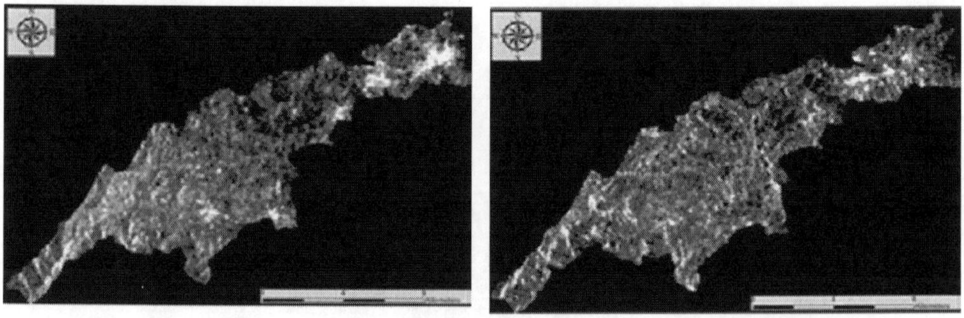

(a) (b)

Figure 12. (a) The catchment area before the precipitation event. (b) The catchment area after the precipitation event

The values of radar backscatter coefficient for the different land cover classes as they were extracted from GeoEye-1 images, are tabulated in Table 3. The results were extracted in GIS environment (ArcGIS 10 software) through the use of zonal statistics application. According to Table 3, the backscatter coefficient in most of the classes increased after the precipitation event. The reason for this phenomenon was the overall moisture increase in the area. The backscatter of forest and urban areas was significantly increased (4.57 and 6.67dB) after the precipitation event due to the double reflection phenomenon. On the other hand, in other classes such as soil and bare rock, dB values declined due to water accumulation and the corresponding surface scattering effect. In agricultural areas of low vegetation, such as alfafa or barley crops, the db were slightly increased.

Table 3. Radar Backscatter of ALOS PALSAR images for different land cover types and days

			Radar Backscatter (dB)	
	Class Name			
		Before Precipitation Event	After Precipitation Event	Difference
1	Bare Rock	-18.83	-24.31	5.48
2	Forest	-23.04	-18.13	4.91
3	Soil	-25.46	-27.94	1.47
4	Trees	-27.94	-22.68	5.26
5	Urban	-23.16	-16.49	6.67
6	Vegetation	-26.84	-26.34	0.50
7	Marl	-31.51	-24.95	6.56

In order to improve image interpretation for water affected areas, several RGB composites were constructed, including microwave and textural bands. The optimum ones improved remarkably the final RGB composites and contributed to the delineation of the moisture affected areas, as shown in Fig. 13. Specifically, in Fig. 13a, the moisture affected areas are indicated in green tones. In Fig. 13b where only texture indicators were used the moisture affected areas are in light cyan color. On the one hand, the combination of speckle reducing Lee filter band and texture indicators in Fig. 13c, resulted in whitish color for flood prone areas. On the other hand, concerning the composite Fig. 13d, the combination of Mean, Var- iance and Homogeneity bands resulted in a light yellowish color for the moisture affected areas.

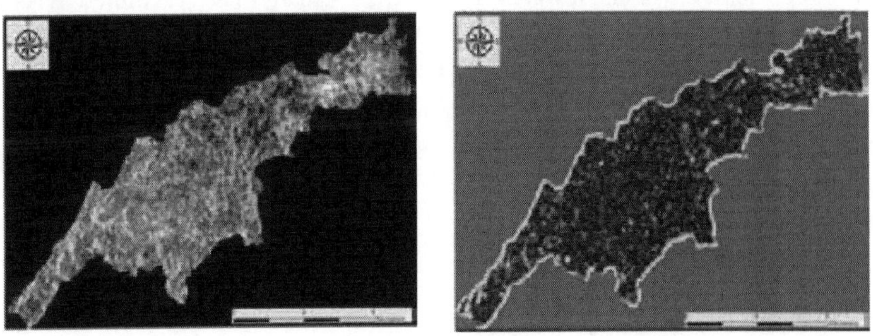

(a) (b)

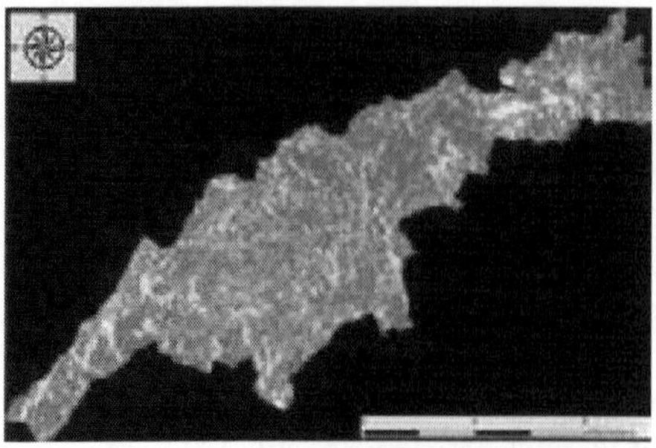

(c)

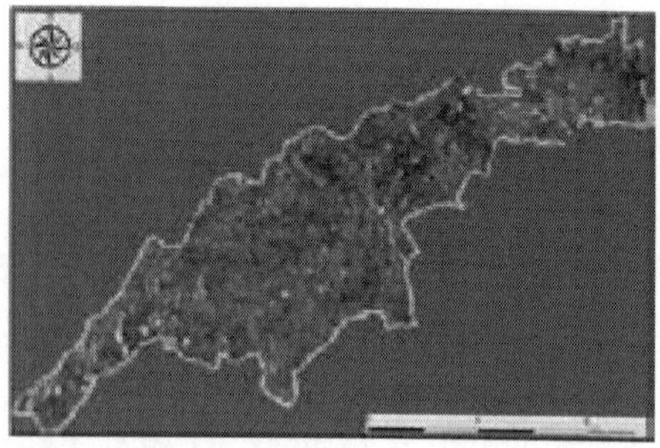

(d)

Figure 13. a) RGB composite of the catchment area with the ALOS images before and after the precipitation event (R: Filtered image before precipitation, G: Filtered image after precipitation, B: Filtered image before precipitation - with green colors the areas where backscattering values were increased due to moisture effect are indicated). (b) Texture indicators RGB composite (R: Homogeneity, G: Contrast, B: Dissimilarity) (c). Combination of microwave bands and textural bands (R: Filtered image before precipitation, G: Filtered image after precipitation, B: Mean). (d) Texture indicators RGB composite (R: Mean, G: Variance, B: Homogeneity)

5.6. Conclusions

In this study, ALOS PALSAR imagery data (acquired before and after a certain precipitation event) proved to be useful for evaluating their potential to detect increased land moisture values and to delineate flood prone areas within a catchment area. In the first approach, signal intensity statistics (backscattering values) were extracted to correlate moisture values with certain land cover classes. For this purpose, two high spatial resolution GeoEye-1 images were used to create a LULC map to be used as a reference thematic map.

In addition, texture analysis was employed to ALOS PALSAR images for the detection of flood prone areas. This method is based on the multi-temporal evaluation of the changes that occur between two ALOS PALSAR overpasses before and after the extreme precipitation event. The specific approach aims to highlight the changes and separate this information from unchanged backscatter signals. Moreover, the specific approach is used in order to improve the visual interpretation of SAR images. The visual inspection of filtered ALOS im- ages proved that there is a considerable change in radar backscattering when moisture affects land cover classes. Relative radar backscatter levels sampled in regions of interest and a LULC cover map indicated that different land cover classes yield different backscatter re- turns in response to moisture/flooding.

The results are useful for examining the potential of ALOS PALSAR images in recording soil moisture regime of an inundated area. However, the research team will continue observation in longer time in case of flooding with the use of radar images. Such information is needed to understand flood mechanism and to better develop water discharge and flood prevention system.

6. Monitoring urban land cover with the use of satellite remote sensing techniques as a means of flood risk assessment in Cyprus.

6.1 Introduction

This study uses an integrated approach that combines record of urban sprawl, land use and landscape metrics. Specifically, a remote sensing approach is applied to Aster satellite im- ages to analyze and identify patterns of urban changes within the spatial limits of Yialias watershed basin in the island of Cyprus. Moreover, there is an effort to optimize the classifi- cation products by combining spectral and texture data to the final.

6.2. Data and methodology

6.2.1. Methodology

An innovative methodology was developed for improving the classification accuracy of Aster images concerning multi-temporal (2000 – 2010) record of urban land cover within the spatial limits of Yialias watershed basin in Cyprus. The phenomenon of spectral similarity of the spectral signatures of urban and marl/chalk formations, identified in the study area, stimulated the calculation of texture measurements in order to improve the traditional classification products derived from spectral bands. Thus, with the use of ENVI 4.7 software 7 indicators of texture information were extracted for the images of 2000 and 2010. These indicators were evaluated for their separability concerning urban and marl / chalk and the optimum ones were used either individually or in combination with spectral bands in order to improve the land use / land cover (LULC) classification accuracy. The Kappa coefficient was used in order to evaluate the reliability of the classified products. In the final stage, the optimum LULC products were incorporated in Fragstats tool in order to record the changes in urban cover structures during the last decade with the use of sophisticated spatial metrics.

6.2.2. Data

For the purposes of the study, the following satellite and digital spatial data were incorporated:

- 2 ASTER Images
- A Digital Elevation Model (DEM) of 25m pixel size provided by the Department of Land and Surveys of Cyprus and created with the use of orthorectified stereopairs of airphotos covering the study area.

The acquired ASTER images have a 10 year time interval in order the multi-temporal moni- toring of urban sprawl to be guaranteed. For this study, the first three spectral bands were used (VNIR and SWIR) with spatial resolution of 15 m. The exact acquisition dates of the images were: 12 May 2000 and 06 April 2010.

6.3. Pre-processing techniques

Geometric corrections were carried out using standard techniques with ground control points and a first order polynomial fit. For this purpose, topographical maps were used to track the position of ground control points in conjunction with the digital shoreline of Cyprus extracted from the provided DEM. in the following, the DN values were converted to radiance values. For both images, the at-satellite radiance values were converted to at–satel- lite reflectance values. Finally, the darkest pixel atmospheric correction method was applied to every image (Hadjimitisis et al., 2004b). It has been found that atmospheric effects con- tribute significantly to the classification technique.

6.4. Image classification

In this study, the Iterative Self-Organizing Data Analysis Technique (ISODATA) method was used. The ISODATA algorithm operates as k-means clustering algorithm by merging the clusters if the separation distance in a multispectral feature is less than a value specified by the user and certain rules for splitting a certain cluster into two clusters. Accuracy assess- ment, which is an integral part of any image classification process, was calculated to esti- mate the accuracy of different methodologies of land cover classifications. An important statistic generated from the error matrix is the Kappa coefficient that is well suited for accu- racy assessment of LULC maps (Vliet, 2009). This statistic takes into account all the values in the matrix and produces an index that indicates the rate of improvement compared to ran- domly allocating pixels to different classes (Congalton & Green, 2008).

The major issue that this study had to deal with was the similarity of spectral signature re- sponse mainly between urban, marl/ chalk and soil features in the Aster images of 2000 and 2010. This problem is clearly denoted in Fig. 14. For this reason different kind of classifica- tion methods were used in order to optimize the final results and provide an alternative way of creating efficient LULC cover maps.

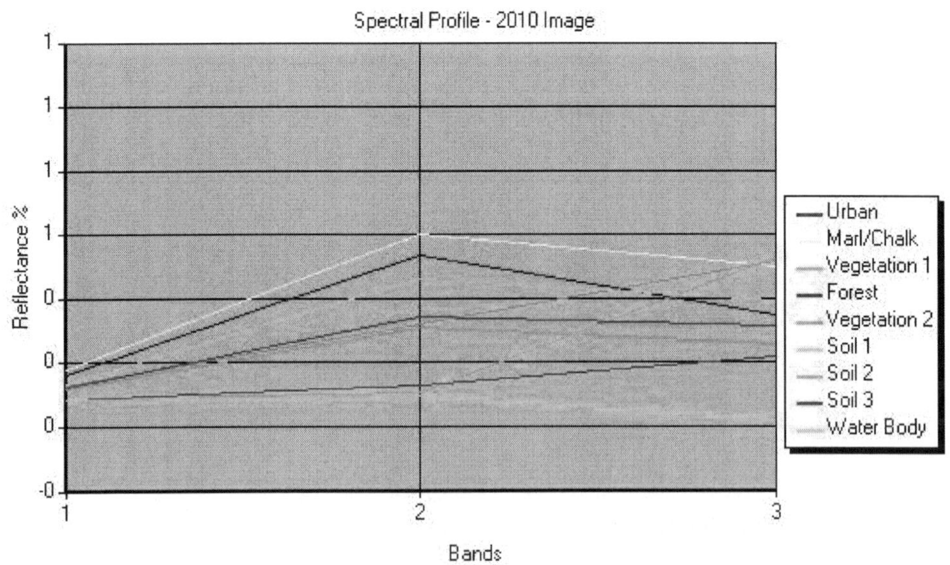

Figure 14. Spectral response curve of typical ground objects

6.4.1. Multispectral image classification

The pixel-based classification is considered to be the most classic way of classifying satellite imagery. For this reason, the first three bands of Aster image were used covering a spectral range from visible to near infrared part of spectrum. This process was accomplished in order to form a standard of comparison with the other classification products such as those of texture or combination of texture and spectral bands. After proceeding with evaluation ac- curacy, it was resulted that the Kappa coefficient for image acquired for 2000 was 0.684 and for 2010 was 0.695. These accuracies can be described as moderate and were ascribed to urban and marl/chalk spectral conflict.

6.4.2. Texture classification

According to Zhang & Zhu (2011), texture is defined as the spatial variation in gray value and is independent of color or luminance. Texture measures smoothness, coarseness and regularity of a region in an image (Gonzalez & Woods, 1992). Concerning satellite digital imagery texture quantifies the way two neighboring pixels relate each other within a small window centered on one of the pixels. It is generally used to describe the visual homogeneity of images and is considered to be a common intrinsic property of

Texture Descriptor	Equation	Description
Contrast	$\sum_{i=0}^{Ng-1}\sum_{j=0}^{Ng-1} (i-j)^2 g^2 (i,j)$	Contrast measures the difference between the highest and lowest values of a contiguous set of pixels. Thus, low contrast image features means low spatial frequencies.
Homogeneity	$\sum_{i=0}^{Ng-1}\sum_{j=0}^{Ng-1} \frac{1}{1+(i+j)2} g(i,j)$	Image homogeneity is sensitive to the presence if near diagonal elements in GLCM.
Entropy	$\sum_{i=0}^{Ng-1}\sum_{j=0}^{Ng-1} g2(i,j)\log(g(i,j))$	Calculates the disorder of an image and gives high values when an image is not texturally uniform
Angular Second Moment (ASM)	$\sum_{i=0}^{Ng-1}\sum_{j=0}^{Ng-1} g(i,j)^2$	ASM measures texture uniformity. High ASM values occur when the distribution of gray levels values is constant.
Dissimilarity	$\sum_{i=0}^{Ng-1}\sum_{j=0}^{Ng-1} g(i,j) \mid i-j \mid$	Dissimilarity is similar to Contrast. However it weights increase linearly rather than weighting the diagonal exponentially.
Mean	$\sum_{i=0}^{Ng-1}\sum_{j=0}^{Ng-1} g(i,j)$	Measure of similarity in pixel values (mean pixel value) of the neighborhood resolution cells in an image block.
Variance	$\sum_{i=0}^{Ng-1}\sum_{j=0}^{Ng-1} (i-u)^2 g(i,j)$	Variance measures homogeneity and increases when the grey level values differ from their mean.

N_g is the number of gray levels, entry (i, j) in the GLCM and $u = \sum_{i=0}^{Ng-1}\sum_{j=0}^{Ng-1} g(i,j)$

Table 4. Description of the texture parameters

all ground objects. For the description of texture histograms, gray level co-occurrence matrix (GLCM), local statistics and characteristics of the frequency spectrum are used. The GCLM mainly operates by calculating a matrix that is based on quantifying the difference between the grey levels of neighboring pixels in an image window. The main aim of this matrix is the quantification of the spatial pixel structure within this window. It was initially suggested as a mechanism for extracting texture measures (Haralick et al., 1973).

Initially, principal component analysis was applied to both satellite images in order to extract the first principal component from each image which would subsequently be used for texture analysis. Thus, the first component of the two images was imported in ENVI 4.7 soft- ware and 7 different statistical indicators of texture such as contrast, angular second moment, homogeneity, entropy, dissimilarity, mean and variance were used for carrying out the statistical texture analysis of all the typical ground objects (Tables 4, 5 and 6).

Table 5. Analysis of texture features of basic objects for satellite image corresponding to 2010

	Contrast	Homogeneity	Entropy	Angular Second Moment	Dissimilarity	Mean	Variance
1 Urban	12.186	0.2841	2.082	0.129	2.779	35.591	4.662
2 Vegetation 1	0.5181	0.778	1.138	0.398	0.455	24.828	0
3 Vegetation 2	2.568	0.560	1.538	0.241	1.123	30.604	0.778
4 Forest	1.083	0.694	1.303	0.318	0.690	19.236	0.220
5 Marl/Chalk	24.808	0.198	2.049	0.137	3.882	49.939	3.759
6 Bare Soil	1.139	0.6605	1.269	0.344	0.755	25.799	0.316

Table 6. Analysis of texture features of basic objects for satellite image corresponding to 2000

	Contrast	Homogeneity	Entropy	Angular Second Moment	Dissimilarity	Mean	Variance
1 Urban	6.856	0.422	1.948	0.151	1.875	28.594	2.606
2 Vegetation 1	3.319	0.533	1.528	0.254	1.284	17.77	0.906
3 Vegetation 2	1.867	0.688	1.398	0.299	0.801	26.178	0.948
4 Forest	0.612	0.723	1.223	0.328	0.562	13.248	0.199
5 Marl/Chalk	7.540	0.380	2.114	0.125	2.057	41.463	4.367
6 Bare Soil	0.337	0.831	0.892	0.485	0.337	18.45	0.160

It is clearly shown in Table 5 that marl formations and urban classes which cannot be differentiated (based on spectral features) vary in the means of contrast, homogeneity, dissimilarity and mean texture regarding the image corresponding to 2010 (Fig. 14).

Concerning the texture bands of 2000 (Table 6) the greatest differences in values between marl and urban classes are indicated at mean and variance texture classes.

Texture-based classification methodologies give the opportunity to end users to extend the traditional-based classifiers by incorporating the texture bands into the multispectral bands, in order to coalesce the spectral and spatial information in the final product. The ISODATA algorithm was applied to different texture products. Specifically, the algorithm was applied to the multiband texture images of 2000 and 2010 and to the PCA products (three first com- ponents) of 2000 and 2010 with corresponding Kappa coefficients of 0.694, 0.685, 0.710, 0.715 and 0.723.

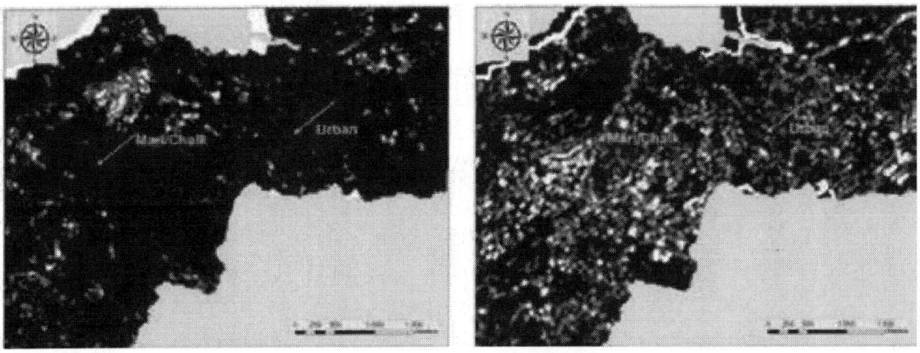

Figure 15. Urban and marl/chalk features as indicated in Angular Second Moment texture indicator (left). Urban and marl/chalk features as indicated in Contrast texture indicator (right)

6.4.3. Combined spectral and texture methodology

The combined use of spectral and texture methodology function by combining spectral and texture bands (either original bands or PCA components) and creating a final integrated image. For this study, the following two combinations were accomplished and the ISODATA classifier was applied to them:

- Use of all multispectral and texture bands
- Use of all multispectral bands and the first three components after the application of PCA to texture bands.

The overall accuracy of the methodology was considered as promising compared to the results of the previous classification products derived from individual either spectral or texture bands. Specifically, the Kappa coefficient values for the 1st category of combined classification for 2000 and 2010 was 0.702 and 0.732, respectively. In addition, the Kappa coefficient values for the second category were 0.765 and 0.775 concerning 2000 and 2010 images. These results led the research team to select these two final LULC cover maps concerning the period 2000 and 2010 for applying spatial landscape metrics.

6.5. Landscape metrics

Spatial landscape metrics are used in sustainable landscape planning and analysis of urban land use change (Botequilha et al., 2002). These metrics typically measure spatial configuration of landscapes, and can be used to enhance the understanding of relationships between spatial pat- terns and spatial processes (Herold et al., 2005). In this study, the FRAGSTATS tool was used in order to measure and analyze the diachronic changes of LULC regime of the study area and re- cord the urban sprawl phenomenon within the watershed. Specifically, seven spatial individual metrics were used for analyzing urban land cover changes and these were (Edge Density, Largest Patch Index, Class Area, Number of Patches, Area weighted mean patch fractal dimension, Euclidean nearest neighbor distance and Contagion) (Table 7).

As investigated by O'Neil et al. (1988) due to correlation and overlap between landscape metrics, it is not necessary to calculate all landscape metrics. The specific metrics were selected because of their simplicity and effectiveness in depicting urban forms evolution (Al- berti & Waddel, 2002; Herold et al., 2002). It was found that there was an increase in built up areas during the period 2000 to 2010. The number of patches used in landscape analysis indicate the aggregation or disaggregation in the landscape. The considerable increase of the specific index during the time span 2000 - 2010 suggests urbanization in the study area characterized by dispersion. Moreover, a development of a number of isolated and fragmented built up areas occurred at the end of this period. Regarding largest patch index, the small increase between 2000 and 2010 indicates a corresponding small urban core increase. The in- creased urbanization rate is characterized by the appearance of new, dispersed settlements.

Table 7. Properties of spatial metrics used in this study

No	Landscape Metrics	Description	Comments
1	Edge Density (ED)	Equals the sum of the lengths of all edge segments divided by total landscape area	It is an absolute measure of total edge length on a. per unit area bases that facilitates comparison. among landscapes of different sizes
2	Largest Patch Index	Equals the area of the largest patch of the corresponding patch type divided by total landscape area and multiplied by 100.	Quantifies the percentage of total landscape area comprised by the largest patch
3	Class Area	Equals the sum of the areas of all patches of the corresponding patch type	Is a measure of landscape composition and calculates how much of the landscape is comprised of a particular landscape.
4	Number of Patches	Equals the number of patches of the corresponding class	Measurement of the extent of subdivision or fragmentation of the patch type.
5	Euclidean Nearest Neighbor Distance	Equals the distance to the nearest neighboring patch of the same type	Simple measure of patch context. It is extensively used for quantification of patch isolation

6	Contagion	Describes the heterogeneity of a landscape	Measures the extent to which landscapes are aggregated or clumped
7	Area weighted mean patch fractal dimension	Area weighted mean value of the fractal dimension values of all the patches	It reflects shape complexity across a range of spatial scales

Thus, the increase of edge density value by indicates an increase in the total length of the edge of the urban patches due to urban land use fragmentation. This finding is also enhanced by the increase in weighted mean patch fractal dimension value indicating the urban sprawl phenomenon in the study area. Moreover, the fractal shape dimension value was always slightly higher than 1, indicating a moderate shape complexity. In addition, the decrease in Euclidean Nearest Neighbor Distance metric between 2000 and 2010 denoted a reduction in the distance between the built-up patches, suggesting coalescence (Table 8).

Table 8. Landscape indices

			Year	
No	**Metrics**	**2000**		**2010**
1	Edge Density	0.7014		2.8892
2	Largest Patch Index	0.0003		0.0005
3	Class Area (km2)	6.042		18.123
4	Number of Patches	1794		7894
5	Euclidean Nearest Neighbour Distance	1886.36		593.2545
6	Contagion	54.845		47.8295
7	Area weighted mean patch fractal dimension	1.0021		1.0061

However, it is important to mention that the landscape metrics results can be used as gener- al indicators and do not provide the users with absolute answers.

6.6. Results

The impacts of changes in land use patterns on hydrology due to extensive urbanization in the spatial limits of watershed is a critical issue in water resource management and watershed land use planning. Land use and land cover maps of the study area for the years 2000 and 2010 were obtained using spectral bands, texture bands or combination of both of them. The major motivation for the use of alternative classification methodologies was the exis- tence of similar spectral signatures for urban and marl/chalk geologic formations located in the study area. These methodologies were evaluated for their accuracy and the optimum classification products were selected in order to be used to the research of urban land use regime evolution during the last decade. In both cases (2000 and 2010) the combination of three spectral bands with the first three principal components extracted from texture bands led to more accurate and reliable results. In the next stage, landscape spatial metrics were used to measure the urban sprawl phenomenon in the study area and its changes through time. Specifically, seven metrics were applied to the two final classified images. The results from the vast majority of the metrics, besides Euclidean distance measurement, denoted a steady dispersion of urban settlements within the area of watershed. Although there was not a significant total urban area increase during this period, a considerable urban sprawl phenomenon was recorded.

This study denoted that spatial measures, such as texture, can play an important role in the analysis of satellite imagery. The overall improvement of classification accuracy products derived from images of medium spatial resolution such as those of Aster highlights the potential of use of texture bands in combination with multispectral imagery. Moreover, the urban sprawl phenomenon was recorded in detail with the use of landscape metrics emphasizing to the flood inundation danger in an already flood prone watershed basin such as Yialias. The research team will continue the specific research by incorporating images of higher spatial resolution to the classification model.

7. Overall conclusions

This study revealed that the integrated use of satellite remote sensing and GIS technology can contribute substantially to the sustainable management of a watershed basin. Interpretation of multi-spectral satellite sensor data proved to be of great help in the development of updated LULC maps and record of the LULC regime and urban sprawl phenomenon in a catchment area. Moreover, a soil erosion model such as RUSLE was found to be efficiently applied at basin scale with quite modest data requirements in a Mediterranean environment. The RUSLE model provides the end users with reliable quantitative and spatial information concerning soil erosion and erosion risk in general. Following, the results denoted the potential of Radar imagery in recording soil moisture regime of an inundated area as well its potential to improve classification accuracy.

The overall results pointed out the substantial contribution of satellite remote sensing to the sustainable management of a catchment area.

References

1. Agapiou, A. & Hadjimitsis, D.G. (2011). Vegetation indices and field spectroradiometric measurements for validation of buried architectural remains: verification under area surveyed with geophysical campaigns, *Journal of Applied Remote Sensing*, Vol. 5, doi:10.1117/1.3645590

2. Alberti, M. & Waddel, P. (2000). An integrated urban development and ecological simulation model, *Integrated Assessment*, Vol 1, pp. 215-227

3. Alexakis, D.D; Hadjimitsis, D.G.; Agapiou, A., Themistocleous K. & Retalis, A. (2011). Contribution of Earth Observation to flood risk assessment in Cyprus: the Yialias catchment area in Nicosia, *Proceedings of VI EWRA International Symposium - Water Engineering and Management in a Changing Environment*, Catania, Italy, June 29 - July 2, 2011

4. Alexakis, D.D.; Hadjimitsis, D.G.; Agapiou, A. & Retalis, A. (2012). Optimizing statistical classification accuracy of satellite remotely sensed imagery for supporting fast flood hydrological analysis, *Acta Geophysica*, Vol 60(3), pp 959-984, doi: 10.2478/s11600-012-0025-9

5. Alexakis, D.D.; Hadjimitsis, D.G. & Agapiou, A. (2013a). Estimating Flash Flood Dis- charge in a Catchment Area with the Use of Hydraulic Model and Terrestrial Laser Scanner, *Advances in Meteorology, Climatology and Atmospheric Physics Springer Atmos- pheric Sciences*, pp 9-14, doi: 10.1007/978-3-642-29172-2_2

6. Alexakis, D.D.; Hadjimitsis, D.G.; Michaelides, S.; Tsanis I.; Retalis, A.; Demetriou, A.; Agapiou A.; Themistocleous K.; Pashiardis S.; Aristeidou, K. & Tymvios F. (2013b). Application of GIS and Remote Sensing Techniques for Flood Risk Assess- ment in Cyprus, *Advances in Meteorology, Climatology and Atmospheric Physics Springer Atmospheric Sciences*, pp. 9-14. doi: 10.1007/978-3-642-29172-2_1

7. Barredo, J. & Engelen, G. (2010). Land Use Scenario Modeling for Flood Risk Mitiga- tion. *Sustainability,* pp.1327-1344; doi:10.3390/su2051327

8. Bou Kheir R; Abdallah, C & Khawlie, M. (2008). Assessing soil erosion in Mediterra- nean karst landscapes of Lebanon using remote sensing and GIS, *Engineering Geology*, Vol 99, pp. 239–254

9. Eiumnoh, A. & Shrestha, R. (2000). Application of DEM data to Landsat image classi- fication: Evaluation in a tropical wet-dry landscape of Thailand, *Photogrammetric En- gineering and Remote Sensing*, Vol 66, pp. 297-1304

10. Ferro, V; Giordano, G. & Lovino, M. (1991). Isoerosivity and erosion risk map for Si- cily, Hydrological Sciences Journal, Vol 36(6), pp.549–564

11. Hadjimitsis, D.G.; Retalis, A. & Clayton, C. (2004a). Satellite remote sensing and GIS for sustainable development in Skiathos Island, Greece, In: *Proceedings SPIE*, Vol.63, 5239 doi:10.1117/12.511522

12. Hadjimitsis, D.G.; Clayton, C.R.I. & Hope, V.S. (2004b). An assessment of the effec- tiveness of atmospheric correction algorithms through the remote sensing of some reservoirs, *International Journal of Remote Sensing*, Vol. 25, pp. 3651-3674

13. Hadjimitsis, D.G. (2007) The use of satellite remote sensing and GIS for assisting flood risk assessment: a case study of the Agriokalamin Catchment area in Paphos- Cyprus, *In: Proceedings SPIE, 6742, 67420Z* ; doi:10.1117/12.751855

14. Hadjimitsis, D.G. (2010). The importance of monitoring urban growth and land-cover changes in catchment areas in Cyprus using multi-temporal remotely sensed data, *Natural Hazards and Earth System Sciences Journal*, Vol.10, pp. 2235-2240, doi:10.5194/ nhess-10-2235-2010

15. Hadjimitsis, D.G., Clayton, C. & Toulios, L., (2010a). Retrieving visibility values us- ing satellite remote sensing data, *Physics and Chemistry of the Earth*, Parts A/B/C, 35 (1–2), pp. 121-124, doi: 10.1016/j.pce.2010.03.002

16. Hadjimitsis, D.G.; Perdikou S. & Themistocleous, K. (2010b). Overview of remote sensing applications for assessing and monitoring natural hazards in Cyprus, *In: Pro- ceedings SPIE*, 7826, 78262B

17. Haralick, R.M.; Shanmugam, K. & Dinstein, I. (1973). Textural features for image classification, *IEEE Transactions on Systems*, Man, and Cybernetics SMC-3, Vol. 3, pp. 610–621

18. Herold, M.; Scepan, J. & Clarke, C. (2002). The use of remote sensing and landscape metrics to describe structures and changes in urban land uses, *Environmental Planning Journal.*, Vol.34, pp. 1443-1458
19. Herold, M.; Couclelis, H. & Clarke, K.C. (2005). The role of spatial metrics in the anal- ysis and modeling of urban land use change, *Computers. Environment and Urban Sys- tems*, Vol.29, pp. 369-399
20. Hongga, Li.; Huang., B. & Xiaoxia Huang, X., (2010). A Level Set Filter for Speckle Reduction in SAR Images. *EURASIP Journal on Advances in Signal Processing*, Vol. 2010, doi : 10.1155/2010/745129
21. Kouli, M.; P Soupios, P. & Vallianatos, F. (2009). Soil erosion prediction using the Revised Universal Soil Loss Equation (RUSLE) in a GIS framework, Chania, Northwestern Crete, Greece, *Environmental Geology* , Vol.57, 483–497
22. Karydas, C.; Sekuloska T. & Silleos, G. (2009). Quantification and site-specification of the support practice factor when mapping soil erosion risk associated with olive plantations in the Mediterranean island of Crete, *Environmental Monitoring and Assessment.*, Vol. 149, pp. 19–28, doi: 10.1007/s10661-008-0179-8
23. Lewis, A.J. (1998). Geomorphic and hydrologic applications of active microwave re- mote sensing in 5 principles and application of imaging radar, *Manual of Remote Sens- ing*, Vol 2, John Wiley & Sons Inc., New York, pp. 567–618
24. Lillesand, T.,M. & Kiefer, R.W. (2000). *Remote Sensing and Image ýnterpretation.* Fourth edition. John Wiley & Sons, Inc., Toronto. ISBN:0-471 25525-7
25. Lin, Y.P.; Lin, Y.B.; Wang, Y.T. & Hong, N.M. (2008). Monitoring and prediction land-use changes and the hydrology of urbanized Paochiao Watershed in Taiwan us- ing remote sensing data urban growth models and a hydrological model, *Sensors*, Vol.8, pp. 680–685
26. Michaelides, S.; Tymvios, F. & Michaelidou, T. (2009). Spatial and temporal characteristics of the annual rainfall frequency distribution in Cyprus, *Atmospheric Research*, Vol.94, pp. 606–615
27. Moore, I. D. & Burch, F.J. (1986). Physical basic of the length–slope factor in the Uni- versal Soil Loss Equation. *Soil Science Society of America Journal,* Vol. 50, pp. 1294–1298
28. Murray, H.; Lucieer, A. & Williams, R. (2010). Texture-based classification of sub-Antarctic vegetation communities on Heard Island, *International Journal of Applied Earth Observation and Geoinformation*, Vol.12, pp. 138–149
29. Nekhay, O.; Arriaza, M. & Boerboom, L. (2009). Evaluation of soil erosion risk using Analytic Network Process and GIS: A case study from Spanish mountain olive plantations, *Journal of Environmental Management*, Vol.90, pp. 3091 – 3104
30. O'Neill, R.V.; Krummel, J.R.; Gardner, R.H.; Sugihara, G.; Jackson, B.; Deangelis, D.L.; Milne, B.T.; Turner, B.T.; Zygmunt, B.; Christensen, S.W.; Dale, V.H. & Graham, R.L. (1988). Indices of landscape pattern, *Landscape Ecology*, Vol.1, pp. 153–162

31. Peijun, D.; Xingli, L.; Wen, C.; Yan, L. & Huapeng, Z. (2010). Monitoring urban land cover and vegetation change by multi-temporal remote sensing information. *Mining Science and Technology*, Vol.20, pp. 0922–0932

32. Prasannakumar, V.; Vijith, H. & Geetha, N. (2011). Estimation of soil erosion risk within a small mountainous sub-watershed in Kerala, India, using Revised Universal Soil Loss Equation (RUSLE) and geo-information technology, *Geoscience Frontiers*, doi:10.1016/j.gsf.2011.11.003

33. Renard, K.G. & Freimund, J.R. (1994). Using monthly precipitation data to estimate the R factor in the revised USLE, Journal of Hydrology, Vol.157, pp. 287–306

34. Rongqun, Z. & Daolin, Z. (2011). Study of land cover classification based on knowledge rules using high-resolution remote sensing images, *Expert Systems with Applications*. Vol.38, pp. 3647–3652

35. Tim, S. & Mallavaram, S. (2003). Application of GIS Technology in Watershed-based Management and Decision Making, *Watershed Update*, Vol.1, pp.1-6

36. Yingxin, Z.; Linlin, G. (2010). Using passive and active remote sensing in combination with GIS for bushfire detection, *In Proceedings 15th Australasian Remote Sensing & Photogrammetry Conference*, Alice Springs, 13-17 September

37. Zhang R. & Zhu D. (2011). Study of land cover classification based on knowledge rules using high-resolution remote sensing images, *Expert Systems with Applications*, Vol.38, pp. 3647-3652

Chapter-5

Remote Sensing for Water Quality Surveillance in Inland Waters: The Case Study of Asprokremmos Dam in Cyprus

1. Introduction

Monitoring, protecting, and improving the quality of water in lakes and reservoirs is critical for targeting conservation efforts and improving the quality of the environment (Ritchie et al., 1994; Nellis et al., 1998). The standard traditional mapping and monitoring techniques of lakes have already become too expensive compared with the information achieved for environ- mental use (Östlund et al., 2001).

Sustainable management of freshwater resources has gained importance at regional (e.g., European Union, 2000) and global scales (United Nations, 2002, 2006; World Water Council, 2006), and 'Integrated Water Resources Management' has become the corresponding scientific paradigm (IPCC, 2007). Water resources, both in terms of quantity and quality, are critically influenced by human activity, including agriculture and land-use change, construction and management of reservoirs, pollutant emissions, and water and wastewater treatment (IPCC, 2008).

Traditional water quality monitoring typically involves costly and time consuming in- situ boat surveys in which in situ measurements or water samples are collected and returned to laboratory for testing of water quality indicators e.g. chlorophyll-a (indicator of algae) and suspended solids. This method allows accurate measurements within a water body but only at discrete points, they can't give the real-time spatial overview that is necessary for the global assessment and monitoring of water quality (Curran et al., 1987; Wang et al., 2004; Brivio et al., 2001).

The challenge of water-quality management associated with the principle of sustainable development has been of concern to many researchers and managers in the last decade. A variety of models have been developed for supporting missions of water-quality management. are becoming more and more important for water-quality management, due to the rapid development of computational problem-solving tools and the enhancement of scientific approaches for information support (Huang & Xia, 2001).

The principal benefit of satellite remote sensing for inland water quality monitoring is the production of synoptic views without the need of costly in-situ sampling. Synoptic, multi- sensor satellite data products and imagery have become increasingly valuable tools for the assessment of water quality in inland and nears-shore coastal waters. Remote sensing of lakes using satellite images has the potential to produce a truly synoptic tool for the monitoring of water quality variables such as chlorophyll a (chl-a), total suspended sediment (TSS), sus- pended minerals (SM), turbidity, Secchi Disk Depth (SDD), particulate organic carbon and coloured dissolved organic matter (CDOM) (Allan et al., 2011; Hadjimitsis, 1999; Mayo et al., 1995; Zhang et al., 2002).

Many researchers have attempted to develop algorithms or models for monitoring water quality in different types of inland water bodies from several satellite sensors such as *Landsat MSS, TM or ETM+ data* (Baban, 1993; Mayo et al., 1995; Östlund et al., 2001; Olmanson, Bauer, & Brezonik, 2008; Lillesand et al., 1983; Wang et al., 2004), *SPOT HVR data* (Lathrop & Lillesand 1989; Chacon-Torres et al., 1992; Jensen et al., 1993; Bhatti, Suttinon, & Nasu, 2011), *MODIS data* (Chen, Hu, & Muller-Karger, 2006; Doxaran et al., 2009; Dall'Olmo et al., 2005), *NOAA AVHRR data* (Prangsma & Roozekrans, 1989; Stumpf & Pennock, 1989; Carrick et al., 1994; Woodruff et al., 1999; Ruhl et al., 2001; Chen et al., 2004), *MERIS data* (Ruiz-Verdú et al., 2008; Guanter et al., 2010; Bresciani et al., 2012), *ASTER data* (Kishino et al., 2005; Nas et al., 2009), *IRS-1C data* (Xu et al., 2010; Sheela et al., 2011), *Hyperion data* (Kutser, 2004; Wang et al., 2005; Giardino et al., 2007), *IKONOS and QuickBird data* (Sawaya et al., 2003; Ekercin, 2007; Oyama et al., 2009). Statistical techniques have been used to investigate the correlation between spectral wavebands or waveband combinations and the desired water quality parameters. Predictive equations for water quality parameters have been developed after these correlations have been determined.

This Chapter describes how remote sensing has been used to monitor water quality in large dams in Cyprus using spectroradiometric measurements and satellite imagery.

2. Monitoring turbidity in dams in Cyprus using remote sensing

2.1. Water quality monitoring in dams

The climate of Cyprus is typical Mediterranean with hot dry summers and mild wet winters, with an average precipitation 500mm per year falling mostly in the winter months. In the last years Cyprus is suffering from water scarcity caused by repeated droughts (Charalambous, 2001; Tsiourtis, 1999; Margat & Vallée, 2000).

The recorded rainfall corresponds to the long-term average of the years 1986 to 2000 gives an average rainfall which is 14% lower than the long-term average of the years 1916 to 1985. In the same period the measured inflow to the existing dams was lower than the previous years' average by 35-40%. Cyprus as a semi-arid country with a highly variable climate, it is predicted that there will be increasing water shortages with the growing water demand in the years to come (Iacovides, 2007; Tsiourtis, 1999). It is important to mention that Cyprus lies heavily on water storage in dams to satisfy its water needs. Today in Cyprus there are 108 dams varying from small ponds to major dams. The fact that the storage capacity of surface reservoirs has reached 304,7 million cubic meters (MCM) of water from a mere 6 MCM in 1960, is a truly impressive achievement when compared to other countries of the same size and level of development as Cyprus.

One of the most important challenges for the Cyprus Water Development Department is the implementation of the European Water Framework Directive (WFD; 2000/60/EC) for inland surface waters including the 108 existing reservoirs. WFD aims at achieving "good water status" and establishes a framework for the protection of all waters including inland surface waters, transitional (estuarine) waters, coastal waters and groundwater by 2015 (Mostert, 2003; Borja et al., 2004). Moreover, WFD sets new objectives for the condition of Europe's water and introduces new means and processes for achieving these objectives. Specific details are also given of the monitoring requirements for different types of water, as well as the assessment and monitoring performance quality standards that should be achieved. For these reasons, monitoring is critical to surface water status within the WFD, as it will determine its classification and the necessity for additional measures in order to achieve the objectives in the Directive (Chen et al., 2004).

Remote sensing technology can become a valuable tool for obtaining information on the processes taking place in the surface of inland water bodies. Satellite remote sensing techniques show more important advantages than traditional sampling with emphasis on the synoptic coverage; it is remarkable that only a single Landsat TM image covers almost all the 108 reservoirs existing in Cyprus; and the high frequency of image captures (Hadjimitsis, 1999; Hadjimitsis et al., 2004a; Hadjimitsis et al., 2010a). Previous studies of using satellite remote sensing in the Cyprus region emphasized the importance of using such techniques for systematic monitoring of water quality either for coastal or inland water bodies due to the good weather conditions in the island (Hadjimitsis et al., 2000; Hadjimitsis et al., 2010a). Moreover remote sensing allows the spatial and temporal assessment of various physical, biological and ecological parameters of water bodies giving the opportunity to examine a large area by applying the suitable algorithm (Hadjimitsis et al., 2006; Hadjimitsis & Clayton, 2009; Papoutsa et al., 2010). Remotely sensed images can give an indication of the physical properties in surface water bodies and can be used to design or improve in-situ sampling monitoring programs by locating appropriate the sampling stations (Dekker et al., 1995). The role of remote sensing technology is there- fore under scrutiny, given its potential capacity for systematic observations at scales ranging from local to global and for the provision of data archives extending back over several decades (Rosenqvist et al., 2003). These issues also highlight a need for the exchange of information between remote sensing scientists and various organizations.

The storage of surface waters in large dams in Cyprus is of vital importance in supplying the local areas for irrigation and potable water supply purposes (Hadjimitsis et al., 2007). At the current time, the Cyprus Water Development Department takes in-situ samples in every dam which provides raw water for treatment, so as to ensure that the water meets the required abstraction standards before it passes to the water treatment works. Previous studies showed the potential of using Landsat TM and Landsat ETM+ remotely sensed data to monitor turbidity in dams in Cyprus. Indeed Hadjimitsis et al., (2007) utilized Landsat TM/ETM+ image data to determine turbidity levels in Kourris Dam, the biggest dam in Cyprus. It is important to mention that turbidity is a vital monitoring parameter for the Water Development Depart- ment, as any high concentrations of suspended solids (i.e more turbid water) may cause serious problems in water filtration processes as shown from other studies (Hadjimitsis, 1999).

Asprokremmos Dam was selected as a case study for the development of a "monitoring tool" for the assessment of Cyprus' inland water quality using remotely sensed data. The concentration of Total Suspended Solids is one of the most critical parameter for the case of Asprokremmos as the water is pumped from the 'outlet area' of the Dam (Deepest point of Asprokremmos Dam / greater distance from the area where Xeros river flows into the Dam) to the water treatment plant of Asprokremmos for pre-treatment and then to the water supply system for the final consumption. High concentrations of suspended particulate matter in reservoir waters directly affect the water treatment plants by occurring damages to the filters during the pretreatment.

During sampling campaigns in Asprokremmos Dam both Turbidity (NTU) and Secchi Disk Depth (SDD) values were determined. Turbidity measures the scattering effect that suspended solids have on light: the higher the intensity of scattered light, the higher the turbidity). Turbidity is measured in Nephelometric turbidity units (NTU) or Formazin turbidity units (FTU), depending on the method and equipment used. Turbidity measured in NTU uses nephelometric methods that depend on passing specific light of a specific wavelength through the sample. FTU is considered comparable in value to NTU and is the unit of measurement when using absorptiometric methods (spectrophotometric equipment) (Wilde & Gibs). SDD is a measure of water clarity by human eyes and all optically active substances in water affect it (Secchi depth decreases as the concentration of chl-a, CDOM, and other substances increas- es). Secchi Disk Transparency is a commonly used, low-cost technique that measures water clarity (Specifically, a black and white disk is lowered into the lake until it can no longer be seen). Water clarity is related to the quantity of phytoplankton in the water, although non- algal turbidity and tannic acids also can reduce water clarity (Fuller et al., 2004).

2.2. Study area

Asprokremmos Dam is built at an altitude of about 100 meters above sea level and is located 16 kilometers east of the city of Paphos. It was completed in 1982 and is the second largest reservoir in Cyprus with a capacity of 52,375,000 cubic meters. It is an earthfill dam, 55 meters high, consisting of the main embankment, spillway, tunnels and galleries and geotechnical works. Due to poor rainfall the dam rarely overflows; the latest times this happened were in 2004 and in 2011. It is considered an important wetland for endemic and migratory birds. The Xeros River that flows into the dam runs only during winter and spring. The study area is shown in Fig 1.

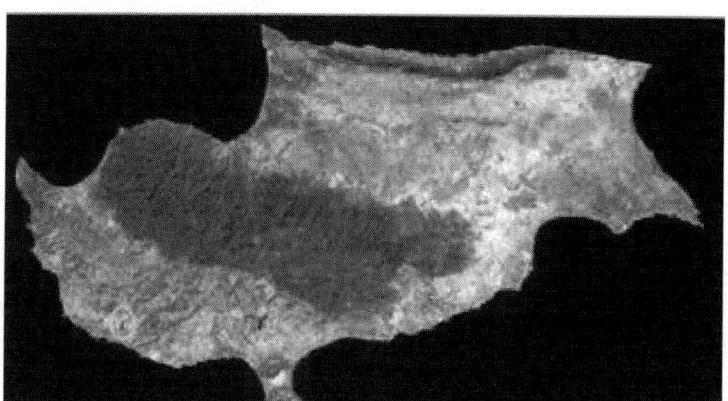

Figure 1. Landsat TM satellite image of Cyprus. Asprokremmos Dam is highlighted

2.3. Pre-processing of satellite images

Pre-processing refers to those operations that precede the main analysis and include mainly geometric and radiometric corrections (Teillet, 1986). Pre-processing steps were applied using the ERDAS Imagine software. All Landsat images were geometrically and radiometrically corrected.

Geometric correction: geometric correction was carried out using standard techniques with ground control points and a first order polynomial fit (Mather, 2004). Twenty well-defined features in the images such as road intersections, airport runway, corners of large buildings were chosen as ground control points (Hadjimitsis et al., 2006).

Radiometric correction: satellite images were converted from digital numbers to units of radiance using standard calibration values. Then the at-satellite radiance values were con- verted to at-satellite reflectance values using the solar irradiance at the top of the atmosphere, Sun-Earth distance correction and solar zenith angle (Mather, 2004). The next step consist the removal of atmospheric effects from satellite imagery. The objective of any atmospheric correction method is to determine the atmospheric effects. Any sensor that records electro- magnetic radiation from the Earth's surface using visible or near-visible radiation will typically register a mixture of two kinds of energy. The value recorded at any pixel location on a remotely sensed image does not represent the true ground-leaving radiance at that point. Part of the brightness is due to the reflectance of the target of interest and the remainder is derived from the brightness of the atmosphere itself. The separation of contributions is not known a priori, so the objective of atmospheric correction is to quantify these two components; in this respect, the analysis can be based on the corrected target reflectance or radiance values. Many atmos- pheric correction methods have been proposed for use with multi-spectral satellite imagery (Hadjimitsis et al., 2004a). In this study, the darkest pixel atmospheric correction method was applied to every image (Hadjimitsis et al., 2004; Hadjimitsis et al., 2010c) since it has been found that DP is a very effective algorithm especially for the visible length. The principle of the DP approach states that most of the signal reaching a satellite sensor from a dark object is contributed by the atmosphere at Visible and Near Infra-Red (NIR) wavelength. Therefore,

the pixels from dark targets are indicators of the amount of upwelling path radiance in this band. The atmospheric path radiance is added to the surface radiance of the dark target, thus giving the target radiance at the sensor.

2.4. Temporal and spatial variations in water quality across the dam

2.4.1. Temporal variations

Eleven (11) Landsat TM/ETM+ satellite images were used in order to investigate how satellite remotely sensed data can become a valuable tool to monitor and assess the *temporal variations* of water quality in Asprokremmos Dam. All the images were pre-processed including geometric and atmospheric correction steps. Atmospheric correction was achieved by applying the Darkest Pixel method for the selected area of Pafos District where Asprokremmos Dam is situated. It has been found from previous studies that the darkest pixel atmospheric correction is the most suitable for inland waters (e.g Hadjimitsis, 1999; Hadjimitsis et al., 2004b). Satellite image processing and analysis was performed using the image processing software (ERDAS Imagine). Table 1 shows the changes of the

Reflectance values observed before and after applying the Darkest Pixel algorithm (ρ_λ % is the reflectance value observed before applying the AC and ρ_{DP} % is the reflectance value observed after applying the AC). It has been shown that an atmospheric correction must be taken into account in the pre-proc- essing of satellite imagery especially where images contain dark targets such as coastal waters or inland waters (Hadjimitsis et al., 2000; 2010b; 2004b; 2009).

Table 1. Mean reflectance values of Landsat Bands 1 to 4 observed in Asprokremmos Dam, before (%) and after (DP %) applying atmospheric correction.

Acquisition Date	Band 1		Band 2		Band 3		Band 4	
	%	DP%	%	DP%	%	DP%•	%	DP%
28-Apr-2004	9.45	1.88	6.79	1.50	4.30	1.15	3.60	1.41
14-May-2004	11.35	1.60	9.03	1.40	7.03	1.78	7.00	3.06
5-Oct-2004	11.42	3.10	8.76	3.11	5.43	1.82	3.67	1.37
13-Aug-2008	14.25	4.46	13.67	5.77	10.68	4.15	6.55	2.29
14-Sep-2008	13.98	3.62	13.08	5.58	10.12	4.42	5.80	1.51
17-Nov-2008	15.07	5.61	14.04	8.08	10.86	7.22	4.83	1.70
29-Jun-2009	9.88	1.37	7.88	1.35	5.14	1.51	3.86	1.31
7-Jul-2009	10.41	1.62	8.54	1.96	5.99	1.17	4.38	1.71
23-Jul-2009	9.82	0.91	8.02	1.04	5.15	1.02	3.38	2.08
1-Sep-2009	11.90	2.37	9.95	2.95	6.87	2.63	4.72	1.93
25-Sep-2009	10.85	3.08	9.92	4.73	6.39	3.80	3.19	2.73

It is obvious in Figure 2 that the maximum reflectance values for Band 2 of Landsat sensor, after applying the atmospheric correction algorithm, are observed in the winter months for

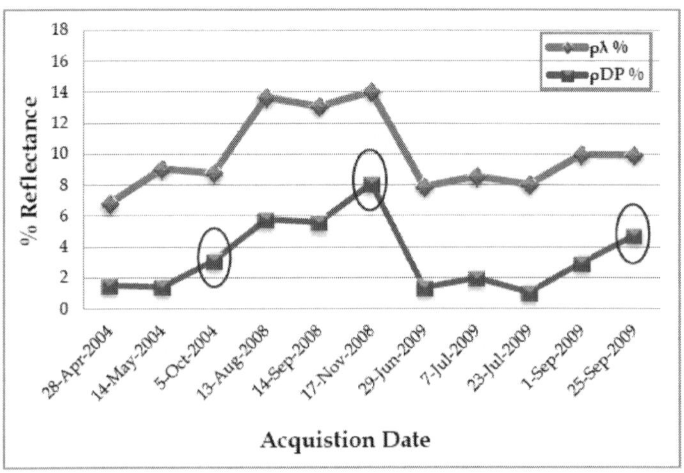

Figure 2. Temporal variations of water quality in Asprokremmos Dam for TM Band 2.

all the years examined (2004, 2008 and 2009). This phenomenon is maybe caused due to the fact that in winter time we have more frequent rain events, and as a result wet deposition of atmospheric particles are observed after each precipitation event.

2.4.2. Spatial variations

Three archived Landsat TM images acquired on 7th July 2009, 23rd July 2009 and 25th September 2009 were analysed in order to assess the *spatial variations* of water quality in the area of Asprokremmos Dam using satellite remotely sensed imagery. The analysis included image pre-processing steps (geometric correction and atmospheric correction) and selection of two areas of the study area (Inlet & Outlet; see Fig. 3) in order to find out the variation of the reflectance in the two areas. The Inlet area is located at the outfall area of the Xeros River where water flows into the Dam and the Outlet area is the area where the water is pumped to the water treatment plant of Asprokremmos for pre-treatment and then to the water supply system for the final consumption. Satellite image processing and analysis were performed using the ERDAS Imagine image processing software and the results are presented in the next section.

The in situ measurements of turbidity have shown that for all the sampling dates the higher values of turbidity were observed for the sampling station which is positioned in the Inlet Area (see Fig. 3) which is where the Xeros River flows into the Asprokremmos Dam while the values reduced along the dam taking the lower turbidity values at the sampling stations which represent the Outlet area. The results of the mean reflectance values of *Inlet & Outlet* areas of Asprokremmos Dam which were calculated in order to find out the variation of the reflectance in the two areas are presented on Tables 2 and 3. These results are in agreement with the in-situ measurements as for all the bands, before and after atmospheric correction the mean reflectance values of the *Inlet areas* are higher than those of the *Outlet areas.*

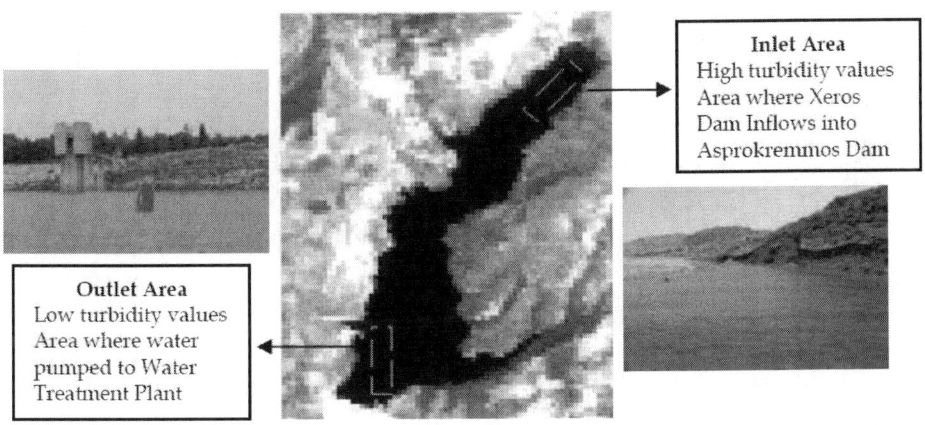

Figure 3. Landsat TM Image focused in Asprokremmos Area pointed the two study areas of the Dam; Inlet & Outlet areas.

Table 2. Spatial variation of mean reflectance values of Inlet & Outlet areas of the Asprokremmos Dam for bands 1 and 2 of Landsat TM multispectral scanning radiometer before and after applying the atmospheric correction.

			Band 1						Band 2			
Acquisition Date		%			DP %			%			DP %	
	Inlet		Outlet	Inlet		Outlet	Inlet		Outlet	Inlet		Outlet
7-Jul-09	11.57		10.43	2.78		1.65	10.76		8.61	4.18		2.03
23-Jul-09	11.15		9.85	2.24		0.94	10.75		8.12	3.76		1.13
25-Sep-09	11.40		10.83	3.64		3.06	10.90		9.84	5.71		4.66

Table 3. Spatial variation of mean reflectance values of Inlet & Outlet areas of the Asprokremmos Dam for bands 3 and 4 of Landsat TM multispectral scanning radiometer before and after applying the DP atmospheric correction.

			Band 3						Band 4			
Acquisition Date		%			DP %			%			DP %	
	Inlet		Outlet	Inlet		Outlet	Inlet		Outlet	Inlet		Outlet
7-Jul-09	8.26		6.03	3.44		1.21	5.34		4.41	2.66		1.74
23-Jul-09	7.78		5.19	3.64		1.06	4.27		3.48	2.96		2.17
25-Sep-09	7.82		6.37	5.24		3.79	4.23		3.29	3.77		2.82

**% is the percentage value of the reflectance before the atmospheric correction and $_{DP}$% is the percentage value of the reflectance after applying the DP atmospheric correction*

2.5 Overall methodology

The overall methodology been applied for the development of an integrated monitoring tool based on remote sensing techniques is briefly presented below (Papoutsa et al., 2011a; 2011b):

- ❖ Design an ideal sampling station network in the area of interest so as to have an adequate number of sampling stations positioned in all directions for the proper and adequate coverage of the study area
- ❖ Collect field spectroradiometric data over the satellite wavelengths during the satellite overpass
- ❖ Retrieve water quality data such as in-situ turbidity & SDD measurements Correlate water quality parameters against spectroradiometric measurements. Retrieve the band with the highest correlation for every inland water quality parameter – extract algorithm
- ❖ Correlate water quality parameters against the at-satellite reflectance after atmospheric correction
- ❖ Use the retrieved equations to monitor the inland water quality parameters

- ❖ Use data collected using the smart buoy for furthermore calibration of the retrieved algorithm due to high frequency of measurements collection (every 2 minutes)
- ❖ Use smart buoy as a monitoring tool able to trigger email or sms alerts when the measure- ments are outside the desired limits

2.6. In-situ turbidity and spectroradiometric measurements

In-situ campaigns in Asprokremmos Dam (Figure 4) were carried out with the collaboration of Cyprus Water Development Department and the Cyprus University of Technology (Remote Sensing Lab) using a power engine boat to collect in-situ data (Figure 5). A sampling station network has been designed in the area of Asprokremmos Dam so as to have an adequate number of sampling stations positioned in all directions for the proper and adequate coverage of the study area and a Global Position System Garmin GPS72 (Figure 6a) was used in order to store and define the preselected sampling stations during the sampling campaigns.

In-situ spectroradiometric data together with in-situ water turbidity readings were collected during the satellite overpass in order to enhance the statistical analyses for retrieving the cross- correlation of spectroradiometric data and water turbidity. A handheld GER-1500 field spectroradiometer (Figure 6b; spectral range covered by the instrument extends from 300 to 1050 nm) equipped with a fibre optic probe was used in order to retrieve the spectral signatures for certain water depths of the Asprokremmos Dam (see Figure 6c). Reflectance was calculated as a ratio of the target radiance to the reference radiance. The target radiance value is the measured value taken 10cm below water surface of the reservoir and the reference radiance value is the measured value taken on the standard Spectralon panel (Figure 6d) representing the sun radiance which reaches the earth surface-without atmospheric influence. The in-situ determination of water turbidity was achieved by using both a portable turbidity meter (Palintest Micro950; Figure 7 b & c) and a Secchi Disk (Figure 7a). Secchi disk depth measure- ments were taken over the shady side of the boat (Papoutsa et al., 2011a; 2011c).

(a)

(b)

Figure 4. Picture of Asprokremmos focused in the (a) Outlet Area & (b) Inlet Area of the Dam.

Figure 5. Power-engine boat with all the required resources which was used during the sampling campaigns in As- prokremmos Dam.

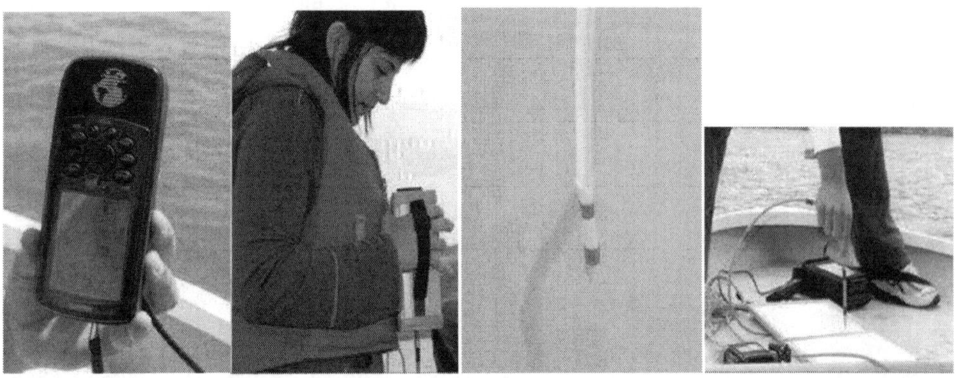

(a) (b) (c) (d)

Figure 6. Equipments used during field campaigns (a) Global Position System Garmin GPS72; (b) handheld GER-1500 field spectroradiometer; (c) fibre optic probe & (d) standard Spectralon panel.

Figure 7. Measuring the turbidity in Asprokremmos Dam using both (a) the Secchi Disk and (b&c) the portable turbidity meter; Palintest Micro950.

2.7. Smart buoy sensor platform

The data buoy (Figures 8a&b) consists of a low cost, low-powered, autonomous floating sensor platform, data logger and gateway to a remote data server. It utilizes an ultra compact powerful embedded system which supports the aggregation of the sensor data, their storage in a local database and transmission of the data to the secure remote storage server. The Data Buoy is highly versatile and can be easily deployed in completely isolated environments for various water monitoring and environmental applications. This robust floating platform can be easily tailored to the specific application needs by selecting different sensors, data logger, powering and communication options.

In our case the buoy has been loaded with various water quality sensors (such as thermometer, turbidity optical sensor - Figure 9a; humidity sensor – Figure 9b etc) and has been deployed in the Asprokremmos Dam, for real time monitoring of water quality (Papoutsa et al., 2011c). The buoy has been used to calibrate the retrieved regression models using the reflectance values as measured at-satellite and turbidity values as measured by the buoy.

(a) (b)

Figure 8. Real time monitoring in Asprokremmos Dam using a Smart buoy sensor platform loaded with various water & environmental quality sensors.

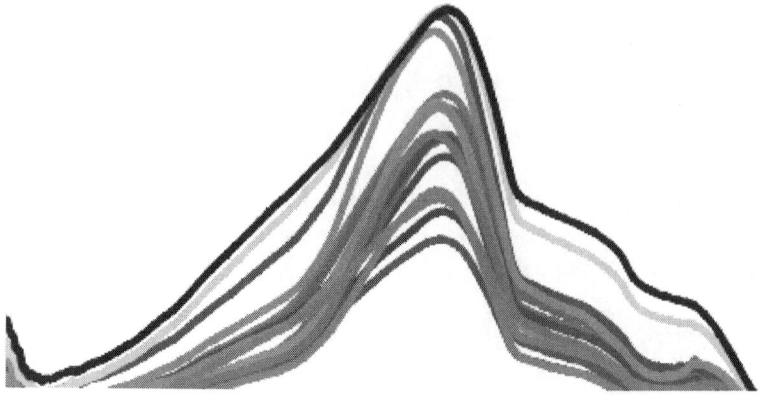

(a) (b)

Figure 9. a) Turbidity optical sensor and (b) humidity sensor located on a floating buoy deployed in Asprokremmos Dam.

3. Results

Spectroradiometric data collected during the field campaigns in two different Areas (Outlet & Inlet Areas of Asprokremmos Dam) characterized by low and high turbidity values are respectively presented in Figures 10a&10b. As we can see examining the typical spectral signatures collected during the in-situ sampling campaigns either the spectroradiometric values or the turbidity values are higher in the Inlet Area of Asprokremmos Dam comparatively to those measured in the Outlet Area of Asprokremmos Dam.

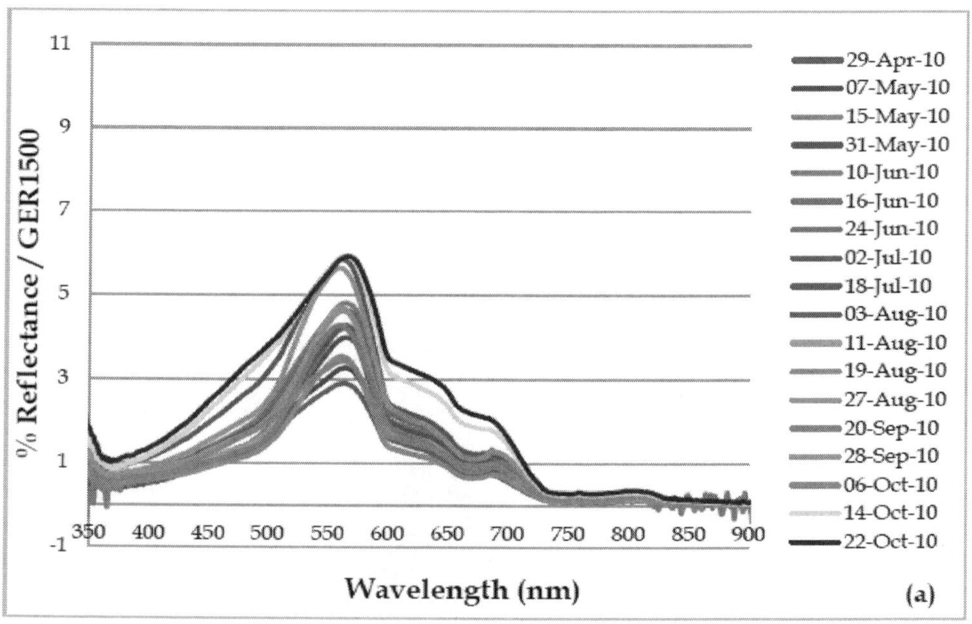

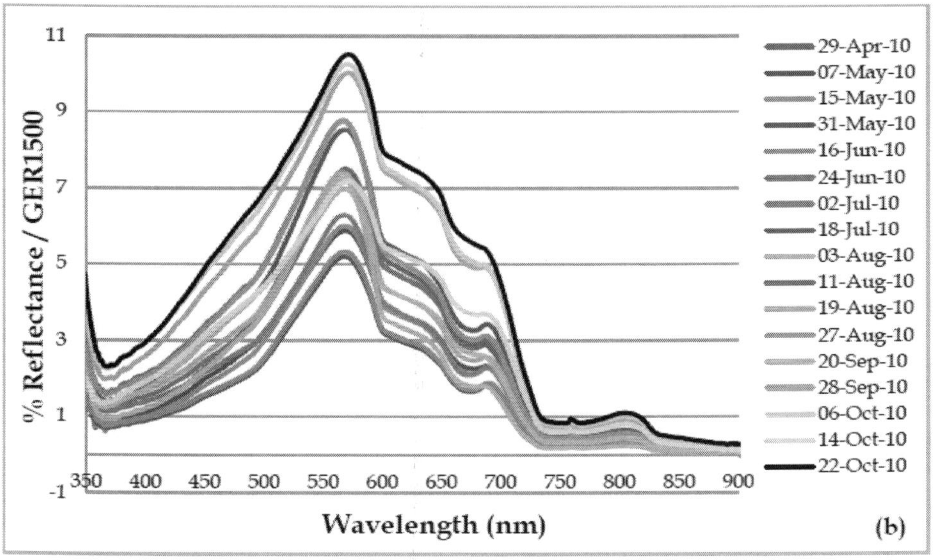

Figure 10. Typical Spectral signatures of (a) Outlet Area & (b) Inlet Area of Asprokremmos Dam acquired using a handheld field spectroradiometer GER1500.

Field spectroradiometric data acquired using the GER1500 provide reflectance data covering the UV, Visible and NIR wavelengths from 350 nm to 1050 nm, with a bandwidth sampling of 1,5nm. All in-situ reflectance data collected using the field spectroradiometer GER1500 were processed inordertogetthemean'in-band'reflectancevaluesforthebands1to4oftheLand satTMandETM + multispectral scanning radiometer and A1 to A62 of the Proba's CHRIS multispectral scanning radiometer. As it can be seen in Figures 11a & 11b Landsat TM has only 4 bands that correspond to the spectral region ranged from 450 to 900 nm against 51 bands (A3 to A53) of Proba / CHRIS and 300 spectral channels of GER1500 field spectroradiometer for the same spectral region.

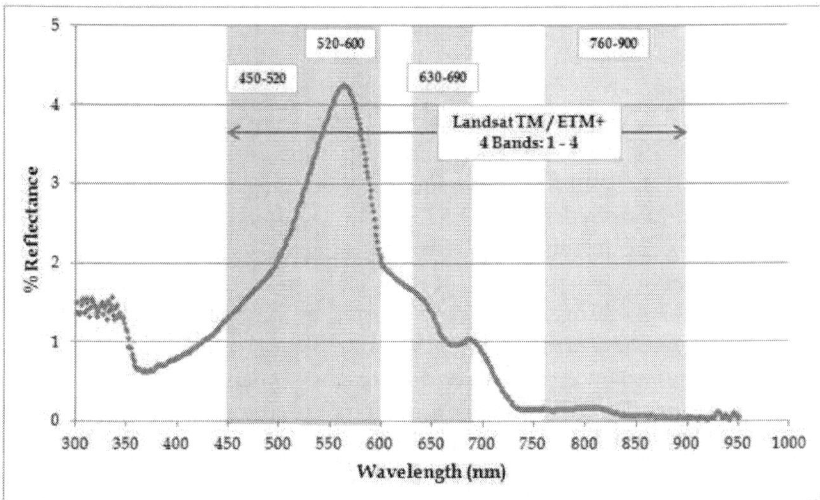

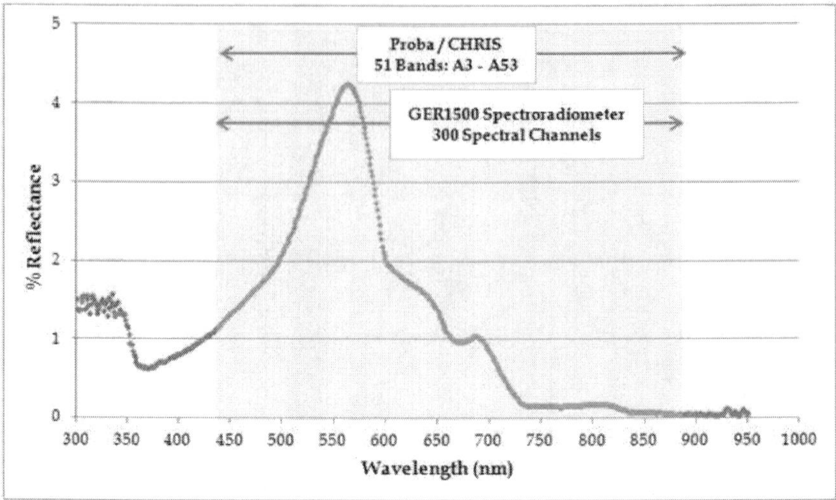

Figure 11. Corresponding Spectral Bands of (a) Landsat TM / ETM+ multispectral scanning radiometer, (b) Proba / CHRIS multispectral scanning radiometer and GER1500 field spectroradiometer of a typical water spectral signature collected during the in-situ sampling campaigns in Asprokremmos Dam – Sampling Station 2.

The mean 'in-band' reflectance value for each sampling point was correlated with the corre- sponding turbidity measurements collected using both the Digital Turbidity-meter and the Secchi Disk for all the bands. This was done in order to identify, the *optimal spectral regions* for monitoring the inland water quality using both Landsat and Proba sensors, and the differences between the two sensors. The methodology adopted in this study is based on the application of linear regression analysis between the mean reflectance values for each band of both the Landsat TM / ETM+ and the Proba / CHRIS (measured with the GER1500 field spectroradi- ometer) across the spectrum and the water turbidity values acquired at the same time at each sampling station in Asprokremmos Dam.

Although, the results depicted better correlation between the in-situ mean reflectance values and the turbidity values taken with the Digital Turbidity Meter than those when the Secchi Disk was used for both sensors (Landsat, Proba) bands. By applying the linear regression model, using the mean in-band reflectance values that correspond to Landsat TM (and ETM +) bands 1-4 as the independent variable and turbidity measurements as the dependent variable for all the combinations, the highest correlation was established between reflectance (acquired from GER1500) in *Landsat TM/ETM+ Band 3 & Band 4*. However, *Band 4* cannot be used for water reflectance measurements because the water absorption coefficient has very high value (near to 1) after 800nm (approximately) and thus, light is mostly absorbed and not reflected by water at wavelengths larger than 800nm. As a result the reflectance at *Band 4* has very low values which can be mostly attributed to measurement errors and despite the apparent high correlation; data corresponding to *Band 4* are not relevant and are not used for the purposes of this study. The very low reflectance values of water at *Band 4* do not give the opportunity for the remote sensing users to retrieve significant aspects regarding water quality. As a result for the determination of turbidity values using Landsat images the optimal band is *Band*

3 with a determination coefficient R^2=0.85 (observed significance level=0.05; equation 1). The same procedure was apply for all 62 bands of Proba / CHRIS A1-A62 and the highest correlation coefficient was found between reflectance (acquired from GER1500) in *Proba / CHRIS Band A31* (Band-width range from 706,2 to 712,4 nm; λmid=709,3 nm) and turbidity with determination coefficient R^2=0.90 (observed significance level=0.05; equation 2) (Papoutsa et al., 2012).

$$y= 0{,}293x + 0{,}387 \quad R^2 = 0{,}85 \tag{1}$$

$$y = 0{,}197x + 0{,}008 \quad R^2 = 0{,}90 \tag{2}$$

Where y values are the mean in-band reflectance values for Landsat ETM+, equation (1) and Proba / CHRIS, equation (2), and x values are the turbidity values measured in *Nephelometric Turbidity Units* (NTU). The values of R^2 indicate the correlation coefficient of the two models (Papoutsa et al., 2012; Papoutsa, C., 2012).

Such outcomes can assist further the remote sensing users for the design of new satellite sensors regarding spectral characteristics for turbidity monitoring campaigns in water dams in the Mediterranean region. Future work consists of calibration and validation campaigns of the proposed regression models based on new satellite imagery acquisitions from both sensors.

The development of regression models such as equation 1 and 2, can be used further to determine turbidity values based on the new satellite acquisitions. Indeed, the authors applied equation 1 using simultaneous measurements both from Landsat ETM+ images and ground truth measurements (spectroradiometric and turbidity). It has been found that the determined turbidity values from satellites after the application of the darkest pixel atmospheric correction, were very close to those found from field campaign. For example, for the Landsat ETM+ image acquired on 31st of May 2010, determined turbidity for an area of interest near the Inlet area was 10,40 NTU (after DP atmospheric correction application) and ground truth turbidity value was 10,04 NTU.

4. Conclusions

Using archived satellite images, spatial and temporal variations of water quality in the Outlet and Intlet areas of Asprokremmos Dam were obvious. Such findings were in accordance with those derived by the in-situ campaigns. It is evident that for all samplings the highest values correspond to the Inlet area where the outfall of the Xeros River exists. It is the area where the water flows into the Dam with a result to carry down clay and suspended solids from the Xeros River resulting in increased values of turbidity accompanied with high reflectance values.

The use of an innovative, energy-autonomous floating sensor platform (buoy) which is installed in the Asprokremmos Dam is used to transfer turbidity data wireless. This can assist further to test and calibrate our developed equation as well as to provide alert to the Cyprus Water Development Department if turbidity values unusually increased.

The use of field spectroscopy assisted the retrieval and definition of the suitable spectral regions that correspond to satellite sensors, such as Landsat TM/ETM+ and Proba /

CHRIS, in which turbidity can be measured and monitored in water dams in Cyprus. Finally the application of atmospheric correction such as the darkest pixel is an essential step prior to any further analysis of satellite imagery.

References

1. Ahern, F. J, Goodenough, D. G, Jain, S. C, & Rao, V. R. (1979). Use of Clear Lakes as Standard Reflectors for Atmospheric Measurements, *In Proceedings of the 11th International Symposium on Remote Sensing of Environment*, Ann Arbor, , 731-755.
2. Allan, M. G, Hamilton, D. P, Hicks, B. J, & Brabyn, L. (2011). Landsat remote sensing of chlorophyll a concentrations in central North Island lakes of New Zealand, *International Journal of Remote Sensing*, , 32, 2037-2055.
3. Baban, S. M. J. (1993). Detecting Water Quality Parameters in Norfolk Broads, UK, Using Landsat Imagery, *International Journal of Remote Sensing*, doi: 10.1080/01431169308953955., 14, 1247-1267.
4. Bhatti, A, Suttinon, P, & Nasu, S. (2011). Monitoring Spatial and Temporal Variability of Suspended Sediment in Indus River by Means of Remotely Sensed Data, *Annals of GIS*, doi:10.1080/19475683.2011.576267., 17, 125-134.
5. Borja, A, Franco, J, Valencia, V, Bald, J, Muxika, I, Belzunce, M. J, & Solaun, O. (2004). Implementation of the European water framework directive from the Basque country (northern Spain): a methodological approach, Marine Pollution Bulletin, doi:10.1016/j.marpolbul.2003.12.001., 48, 209-218.
6. Bresciani, M, Vascellari, M, Giardino, C, & Matta, E. (2012). Remote Sensing Supports the Definition of the Water Quality Status of Lake Omodeo (Italy), *European Journal of Remote Sensing*, doi:10.5721/EuJRS20124530., 45, 349-360.
7. Brivio, P. A, Giardino, C, & Zilioli, E. (2001). Validation of satellite data for quality assurance in lake monitoring applications, *Science of the Total Environment*, , 268, 3-13.
8. Carrick, H. J, Worth, D, & Marshall, M. L. (1994). The Influence of Water Circulation on Chlorophyll-turbidity Relationships in Lake Okeechobee as Determined by Remote Sensing, *Journal of Plankton Research*, doi:10.1093/plankt/16.9.1117., 16, 1117-1135.
9. Chacon-torres, A, Ross, L. G, Beveridge, M. C. M, & Watson, A. I. (1992). The Application of SPOT Multispectral Imagery for the Assessment of Water Quality in Lake Patzcuaro, Mexico, *International Journal of Remote Sensing*, , 13, 587-603.
10. Charalambous, C. N. (2001). Water management under drought conditions, Desalination, , 138, 3-6.
11. Chen, Q, Zhang, Y, Ekroos, A, & Hallikainen, M. (2004). The role of remote sensing technology in the EU water framework directive (WFD), *Environmental Science & Poli- cy*, doi:org/10.1016/j.envsci.2004.05.002., 7, 267-276.
12. Chen, Z, Hu, C, & Muller-karger, F. (2006). Monitoring Turbidity in Tampa Bay Using MODIS/Aqua 250-m Imagery, *Remote Sensing of Environment*, , 109, 207-220.

13. Curran, P. J, Hansom, J. D, Plmnmer, S. E, & Pedley, M. I. (1987). Multispectral remote sensing of nearshore suspended sediments: a pilot study, *International journal of Remote Sensing*, , 8, 103-112.

14. Dall'OlmoG., Gitelson, A. A., Rundquist, D. C., Leavitt, B., Barrow, T. & Holz, J. C., ((2005). Assessing the Potential of SeaWiFS and MODIS for Estimating Chlorophyll Concentration in Turbid Productive Waters Using Red and Near-infrared Bands, *Remote Sensing of Environment*, doi:10.1016/j.rse.2005.02.007., 96, 176-187.

15. Dekker, A. G, Malthus, T. J, & Hoogenboom, H. J. (1995). The remote sensing of inland water quality, *Advances in Environmental Remote Sensing*, edited by Danson, F. M. & Plummer, S. E. (Chichester: John Wiley & Sons), , 123-142.

16. Doxaran, D, Froidefond, J. M, Castaing, P, & Babin, M. (2009). Dynamics of the Turbidity Maximum Zone in a Macrotidal Estuary (the Gironde, France): Observations from Field and MODIS Satellite Data, *Estuarine, Coastal and Shelf Science*, doi:10.1016/j.ecss.2008.11.013., 81, 321-332.

17. Ekercin, S. (2007). Water Quality Retrievals from High Resolution Ikonos Multispectral Imagery: A case Study in Istanbul, Turkey, *Water, Air and Soil Pollution*, doi: 10.1007/s11270-007-9373-5., 183, 239-251.

18. Fuller, L. M, Aichele, S. S, & Minnerick, R. J. (2004). Predicting Water Quality by Re- lating Secchi-Disk Transparency and Chlorophyll a Measurements to Satellite Image- ry for Michigan Inland Lakes, August 2002, U.S. Geological Survey Scientific Investigations Report p., 2004-5086.

19. Giardino, C, Brando, V. E, Dekker, A. G, Strömbeck, N, & Candiani, G. (2007). Assessment of Water Quality in Lake Garda (Italy) Using Hyperion, *Remote Sensing of Environment*, doi:10.1016/j.rse.2006.12.017., 109, 183-195.

20. Guanter, L, Ruiz-verdú, A, Odermatt, D, Giardino, C, Simis, S, Estellés, V, Heege, T, Domínguez-gómez, J. A, & Moreno, J. (2010). Atmospheric Correction of ENVISAT/MERIS Data over Inland Waters: Validation for European lakes, *Remote Sensing of En- vironment*, doi:10.1016/j.rse.2009.10.004., 114, 467-480.

21. Hadjimitsis, D. G. (1999). The application of atmospheric correction algorithms in the satellite remote sensing of reservoirs. PhD thesis. University of Surrey, School of En- gineering in the Environment, Department of Civil Engineering, Guildford (UK).

22. Hadjimitsis, D. G, Clayton, C. R. I, & Hope, V. S. (2000). The Importance of Accounting for Atmospheric Effects in Satellite Remote Sensing: A Case Study from the Lower Thames Valley Area, UK. Proceedings of Space 2000: The Seventh International Conference and Exposition on Engineering, Construction, Operations, and Business in Space, 0-78440-479-8Conf. Proc. (2000). doi:10.1061/40479(204)19, 194-201.

23. Hadjimitsis, D. G, Clayton, C. R. I, & Retalis, A. (2003). Darkest Pixel Atmospheric Correction Algorithm: a Revised Procedure for Environmental Applications of Satellite Remotely Sensed Imagery, *In Proceedings 10th International Symposium on Remote Sensing*, 8-12/9/2003, Barcelona, organised by NASA, SPIE Conference, , 414.

24. Hadjimitsis, D. G, Clayton, C. R. I, & Hope, V. S. (2004a). An assessment of the effectiveness of atmospheric correction algorithms through the remote sensing of some

reservoirs, *International Journal of Remote Sensing*, doi:10.1080/01431160310001647993., 25, 3651-3674.

25. Hadjimitsis, D. G, Clayton, C. R. I, & Retalis, A. (2004b). Darkest pixel atmospheric correction algorithm: a revised procedure for environmental applications of satellite remotely sensed imagery, Proceedings 10th International Symposium on Remote Sensing, 8-12/9/2003, Barcelona- SPAIN, organized by NASA, SPIE CONFERENCE, doi:10.1117/12.511520., 414.
26. Hadjimitsis, D. G, Hadjimitsis, M. G, Clayton, C. R. I, & Clarke, B. (2006). Determination of turbidity in Kourris Dam in Cyprus utilizing Landsat TM remotely sensed da- ta, *Water Resources Management: An International Journal,* doi:10.1007/s11269-006-3089- y., 20, 449-465.
27. Hadjimitsis, D. G, Toulios, L, & Clayton, C. R. I. (2007). Contributions of Satellite Re- mote Sensing to Analysis of Spatial and Temporal Variability of Water Quality in In- land Water Bodies, 10th International conference on Environmental science and technology- 2007 CEST, Kos, Greece, Sept. 2007, , 234-240.
28. Hadjimitsis, D. G, & Clayton, C. R. I. (2008). The Use of an Improved Atmospheric Correction Algorithm for Removing Atmospheric Effects from Remotely Sensed Images using an Atmosphere-surface Simulation and Meteorological Data, Meteorological Applications. doi:met.80
29. Hadjimitsis, D. G, & Clayton, C. R. I. (2009). Assessment of temporal variations of water quality in inland water bodies using atmospheric corrected satellite remotely sensed image data, *Journal of Environmental Monitoring and Assessment*, doi:10.1007/s10661-008-0629., 159, 281-292.
30. Hadjimitsis, D. G, Clayton, C. R. I, & Toulios, L. (2010a). A New Method for Assessing the Trophic State of Large Dams in Cyprus Using Satellite Remotely Sensed Data, *Water and Environment Journal*, doi:10.1111/j.1747-6593.2009.00176.x., 24, 200-207.
31. Hadjimitsis, D. G, Hadjimitsis, M. G, Toulios, L, & Clayton, C. R. I. (2010b). Use of space technology for assisting water quality assessment and monitoring of inland water bodies, *Journal of Physics and Chemistry of the Earth,* doi:10.1016/j.pce. 2010.03.033., 35, 115-120.
32. Hadjimitsis, D. G, Papadavid, G, Agapiou, A, Themistocleous, K, Hadjimitsis, M. G, Retalis, A, Michaelides, S, Chrysoulakis, N, Toulios, L, & Clayton, C. R. I. (2010c). At- mospheric Correction for Satellite Remotely Sensed Data Intended for Agricultural Applications: Impact on Vegetation Indices, *Natural Hazards and Earth System Sciences*, www.nat-hazards-earth-syst-sci.net/10/89/2010/, 10, 89-95.
33. Huang, G. H, & Xia, J. (2001). Barriers to sustainable water-quality management, *Journal of Environmental Management*, doi:10.1006/jema.2000.0394, 61, 1-23.
34. Jensen, J. R, Narumanlani, S, Weatherbee, O, & Mackey, H. E. Jr., ((1993). Measurement of Seasonal and Yearly Cattail and Waterlily Changes Using Multidate SPOT Panchromatic Data, *Photogrammetric Engineering and Remote Sensing*, , 52, 31-36.
35. Iacovides, I. (2007). National Report on «Monitoring progress and promotion of wa- ter demand management policies in Cyprus», Under the aegis of the Water Develop- ment Department and Plan Bleu, Nicosia, Cyprus.

36. IPCC(2007). Climate Change 2007: Freshwater resources and their management. Impacts, Adaptation and Vulnerability. Contribution of Working Group II to the Fourth Assessment Report of the Intergovernmental Panel on Climate Change, Kundzewicz, Z.W., Mata, L.J., Arnell, N.W., Döll, P., Kabat, P., Jiménez, B., Miller, K. A., Oki, T., Sen, Z. & Shiklomanov, I.A. (eds.). Cambridge University Press, Cambridge, UK, , 173-210.

37. IPCC(2008). Climate Change and Water. Technical Paper of the Intergovernmental Panel on Climate Change, Bates, B.C., Kundzewicz, Z. W., Wu, S. & Palutikof, J. P. (eds.). Secretariat, Geneva, , 210.

38. Kishino, M, Tanaka, A, & Ishizaka, J. (2005). Retrieval of Chlorophyll a, Suspended Solids, and Colored Dissolved Organic Matter in Tokyo Bay Using ASTER Data, *Remote Sensing of Environment*, doi:10.1016/j.rse.2005.05.016., 99, 66-74.

39. Kutser, T. (2004). Quantitative Detection of Chlorophyll in Cyanobacterial Blooms by Satellite Remote Sensing, *Limnology and Oceanography*, , 49, 2179-2189.

40. Lathrop, R. G. Lillesand, Jr. & T. M., ((1989). Monitoring Water Quality and River Plume Transport in Green Bay, Lake Michigan with SPOT-1 Imagery, *Photogrammetric Engineering and Remote Sensing*, , 55, 349-354.

41. Lillesand, T. M, Johnson, W. L, Deuell, R. L, Lindstrom, O. M, & Meisner, D. E. (1983). Use of Landsat Data to Predict the Trophic State of Minnesota Lakes, *Photo- grammetric Engineering and Remote Sensing*, , 49, 219-229.

42. Margat, J, & Vallée, D. (2000). Mediterranean Vision on water, population and the environment for the 21st Century, Blue Plan for the Global Water Partnership/Medtac.

43. Mather, P. (2004). Computer Processing of Remotely-sensed Images: An Introduction, 3rd Edition Chichester: Wiley.

44. Mayo, M, Gitelson, A, & Ben-avraham, Z. (1995). Chlorophyll distribution in Lake Kinneret determined from Landsat Thematic Mapper data, *International Journal of Re- mote Sensing*, , 16, 175-182.

45. Moran, M. S, Jackson, R. D, Slater, P. N, & Teillet, P. M. (1992). Evaluation of Simpli- fied Procedures for Retrieval of Land Surface Reflectance Factors from Satellite Sen- sor Output, *Remote Sensing of the Environment*, , 41, 169-184.

46. Mostert, E. (2003). The European Water Framework Directive and water management research, *Physics and Chemistry of the Earth*, Parts A/B/C, doi:10.1016/S1474-7065(03)00089-5., 28, 523-527.

47. Nas, B, Karabork, H, Ekercin, S, & Berktay, A. (2009). Mapping Chlorophyll-a Through In-situ Measurements and Terra ASTER Satellite Data, *Environment Moni- toring Assessment*, doi:10.1007/s10661-008-0542-9., 157, 375-382.

48. Nellis, M. D, & Harrington, J. A. Jr. & Wu, J. ((1998). Remote sensing of temporal and spatial variations in pool size, suspended sediment, turbidity, and Secchi depth in Tuttle Creek Reservoir, Kansas: 1993, Geomorphology Papers, , 21, 281-293.

49. Olmanson, L. G, Bauer, M. E, & Brezonik, P. L. (2008). A 20-year Landsat Water Clarity Census of Minnesota's 10,000 Lakes, *Remote Sensing of Environment*, doi: 10.1016/j.rse.2007.12.013., 112, 4086-4097.

50. Östlund, C, Flink, P, Strömbeck, N, Pierson, D, & Lindell, T. (2001). Mapping of the water quality of Lake Erken, Sweden, from Imaging Spectrometry and Landsat Thematic Mapper, *The Science of the Total Environment*, , 268, 139-154.

51. Oyama, Y, Matsushita, B, Fukushima, T, Matsushige, K, & Imai, A. (2009). Application of Spectral Decomposition Algorithm for Mapping Water Quality in a Turbid Lake (Lake Kasumigaura, Japan) from Landsat TM Data, *ISPRS Journal of Photogrammetric and Remote Sensing*, doi:10.1016/j.isprsjprs.2008.04.005., 64, 73-85.

52. PapoutsaChr., Hadjimitsis, D. G., Themistocleous, K., Perdikou, P., Retalis, A. & Toulios, L., ((2010). Smart monitoring of water quality in Asprokremmos Dam in Paphos, Cyprus using satellite remote sensing and wireless sensor platform, Proc. SPIE 7831, 78310Q. doi:10.1117/12.864824.

53. Papoutsa, C, Hadjimitsis, D. G, & Alexakis, D. D. (2011a). Characterizing the spectral signatures and optical properties of dams in Cyprus using field spectroradiometric measurements, Proceedings SPIE Remote Sensing 2011, Prague, Czech Republic (2011). doi:10.1117/12.898353, 8174

54. Papoutsa, C, Hadjimitsis, D. G, & Alexakis, D. D. (2011b). Coastal water quality near to desalination project in Cyprus using Earth observation. Proceedings SPIE Remote Sensing 2011, Prague, Czech Republic (2011). doi:10.1117/12.898361, 8181

55. Papoutsa, C, Hadjimitsis, D. G, Kounoudes, T, Toulios, L, Retalis, A, & Kyrou, K. (2011c). Monitoring turbidity in Asprokremmos dam in Cyprus using earth observation and smart buoy platform, VI EWRA International Symposium, Water Engineering and Management in a Changing Environment, Catania, Italy- JUNE 29th- JULY 2nd, (2011).

56. Papoutsa, C, Hadjimitsis, D. G, Retalis, A, & Toulios, L. (2012). Defining the LANDSAT TM/ETM+ and CHRIS / Proba spectral regions in which turbidity can be re- trieved in the Asprokremmos Dam in Paphos, Cyprus: Spectro-radiometric measurements campaign *International Journal of Remote Sensing*, Submited., 2010-2011.

57. Papoutsa, C. (2012). PhD thesis Data, Cyprus University of Technology, Remote Sensing Lab, Lemesos

58. Prangsma, G. J, & Roozekrans, J. N. (1989). Using NOAA AVHRR Imagery in Assess- ing Water Quality Parameters, *International Journal of Remote Sensing*, doi: 10.1080/01431168908903921., 10, 811-818.

59. Ritchie, J. C, Schiebe, F. R, Cooper, C, & Harrington, J. A. Jr., ((1994). Chlorophyll measurements in the presence of suspended sediment using broad band spectral sensors aboard satellites, *Journal of Freshwater Ecology*, , 9, 197-206.

60. Rosenqvist, Å, Milne, A, Lucas, R, Imhoff, M, & Dobsone, C. (2003). A review of remote sensing technology in support of the Kyoto Protocol, *Environmental Science & Policy*, doi:10.1016/S1462-9011(03)00070-4, 6, 441-455.

61. Ruhl, C. A, Schoellhamer, D. H, Stumpf, R. P, & Lindsay, C. L. (2001). Combined Use of Remote Sensing and Continuous Monitoring to Analyse the Variability of Sus- pended-Sediment Concentrations in San Francisco Bay, California, *Estuarine, Coastal and Shelf Science*, doi:10.1006/ecss.2000.0730., 53, 801-812.

62. Ruiz-verdú, A, Koponen, S, Heege, T, Doerffer, R, Brockmann, C, Kallio, K, Pyhälahti, T, et al. (2008). Development of MERIS Lake Water Algorithms: Validation Results from Europe, In ESA/ESRIN (Ed.), Proceedings of the 2nd MERIS/AATSR workshop, Frascati, Italy, September URL:http://dx.doi.org/10.5167/uzh-4494., 22-26.

63. Sawaya, K. E, Olmanson, L. G, Heinert, N. J, Brezonik, P. L, & Bauer, M. E. (2003). Extending Satellite Remote Sensing to Local Scales: Land and Water Resource Monitoring Using High-Resolution Imagery, *Remote Sensing of Environment*, doi:10.1016/j.rse.2003.04.0006., 88, 144-156.

64. Sheela, A. M, Letha, J, Joseph, S, Ramachandran, K. K, & Sanalkumar, S. P. (2011). Trophic State Index of a Lake System Using IRS (P6-LISS III) Satellite Imagery, *Environmental Monitoring Assessment*, doi:10.1007/s10661-010-1658-2., 177, 575-592.

65. Stumpf, R, & Pennock, J. (1989). Calibration of a General Optical Equation for Remote Sensing of Suspended Sediments in a Moderately Turbid Estuary, *Journal of Ge- ophysics Research*, C10), , 94, 14.

66. Teillet, P. M. (1986). Image Correction for Radiometric Effects in Remote Sensing, *In- ternational Journal of Remote Sensing*, , 7, 1637-1651.

67. Tsiourtis, N. (1999). Framework for action Mediterranean islands, Global Water Partnership- Mediterranean.

68. Vermote, E. (1996). Atmospheric Correction Algorithm: Spectral Reflectances (MOD09). Algorithm Technical Background Document. NASA , 5-96062.

69. Wang, Y, Xia, H, Fu, J, & Sheng, G. (2004). Water quality change in reservoirs of Shenzhen, China: detection using LANDSAT/TM data, *Science of the Total Environment*, , 328, 195-206.

70. Wang, S, Yan, F, Zhou, Y, Zhu, L, Wang, L, & Jiao, Y. (2005). Water Quality Monitoring Using Hyperspectral Remote Sensing Data in Taihu Lake China. Geoscience and Remote Sensing Symposium, 2005. IGARSS'05. Proceedings. 2005 IEEE International, 25-29 July, doi:10.1109/IGARSS.2005.1526679., 4553-4556.

71. WFD; Directive 2000/60/EC of the European Parliament and of the Council of 23 Oc- tober (2000). Official Journal of the European Communities. (http://www.europa.eu.int/eur-lex/)

72. Wilde, F. D, & Gibs, J. variously dated). 6.7 TURBIDITY, U.S. Geological Survey TWRI Book 9, http://water.usgs.gov/owq/FieldManual/Chapter6/Section6.7.pdf, 1-30.

73. Woodruff, D. L, Stumpf, R. P, Scope, J. A, & Paerl, H. W. (1999). Remote Estimation of Water Clarity in Optically Complex Estuarine Waters, *Remote Sensing of Environment*, doi:10.1016/S0034-4257(98)00108-4., 68, 41-52.

74. Xu, J. P, Li, F, Zhang, B, Gu, X. F, & Yu, T. (2010). Remote Chlorophyll-a Retrieval in Case-II Waters Using an Improved Model and IRS-P6 Satellite Data, *International Journal of Remote Sensing*, doi:10.1080/01431161.2010.485136., 31, 4609-4623.

75. Zhang, Y, Pulliainen, J, Koponen, S, & Hallikainen, M. (2002). Application of an em- pirical neural network to surface water quality estimation in the Gulf of Finland us- ing combined optical data and microwave data, *Remote Sensing of Environment*, , 81, 327-336.

Chapter-6

Detection of Water Pipes and Leakages in Rural Water Supply Networks Using Remote Sensing Techniques.

1. Introduction

Water leakages have been a major problem for many regions around the world (Weifeng et al. 2011). However, monitoring such leakages is a difficult task since traditional field survey methods are costly and time consuming (Huang et al. 2010). Researchers from diverse scientific fields have studied this problem through the development of several techniques including radar technique, geophones, gas filling, and many others. Different conventional techniques such as acoustics, radioactive, electromagnetic, ground penetrating radar and linear polariza- tion resistance have been used over the years for water pipeline leakage detection (Skolnik, 1990; Heathcote and Nicholas, 1998; Hunaidi and Giamou, 1998; Eyuboglu et al., 2003; Burn et al., 2001; Hadjimitsis, et al., 2009).

Remote sensing has been used for a wide range of applications including water management. Studies have shown promising results from its use for water leakage detection (Sheikh Naimullah, 2007). The uses of remote sensing techniques for water leakage detection are time and cost effective compared with traditional, intrusive methods, but their use is restricted due to their spatial resolution. The pipeline leakages occur along the length of the pipeline and the area affected may not be detectable by the satellite sensor as it depends on the pixel size and the density of the vegetation developed due to the presence of water.

Vegetation indices (VI) are the main form of satellite spectral data used for several applications. According to Agapiou *et al.* (2012a), VIs can be divided into five main categories according to equation or the use of each index, which include broadband indices, narrowband indices (hyperspectral), leaf pigment indices, stress indices and water stress indices. They reported that VI can be simply divided according to the wavelength characteristics used in their formula (broadband and narrowband indices). Using airborne remotely sensed imagery, Pickerill and Malthus (1998) analyzed two known water leaks and found that different vegetation indices and single bands were required in order to identify each leak. The spectral profile of one leak responded best to a ratio of NIR to red reflectance, while in the other, NIR to red reflectance ratio was not useful in differentiating it from its surroundings.

Huang et al. (2009) used airborne multispectral remote sensing imagery with high-resolution imaging sensors in the visible, NIR and thermal infrared wavelengths and found that airborne multispectral imaging is a useful tool in the detection of irrigation canal leakage in distribution networks. They concluded that the analysis of the processed image data from red, NIR and thermal bands is highly consistent with the observations from field investigation. Images from individual bands, particularly from the thermal band, can help detect leakage from irrigation canals. The NDVI image, which combines the data from the red and the NIR bands, can help detect and correct errors observed on the thermal imagery.

On-site observation, which consists of data collection, periodical observations, and multivari- ate risk assessment analysis, is the most common technique of monitoring the water pipe network in Cyprus. However, this is difficult to accomplish with traditional methods since it is time consuming, expensive and monitoring is localized. Furthermore, part of the water network tends to be located in inaccessible areas, away from the main road network and urban areas. A complete geoinformation system providing the exact location, characteristics and relevant data for the water mains does not exist, making the leakage monitoring procedures even more challenging.

This paper presents the results from a project which combines different remote sensing technologies for the detection and monitoring of water leakages for water utility systems located in open fields in Cyprus. Two case studies areas were evaluated using freely distributed Landsat 7 ETM+ satellite images and ground spectroradiometric data. In addition, a low altitude system was deployed to observe these pipelines from different heights.

Finally, different remote sensing techniques have been used evaluated as in the detection of leakage from a major water pipe in Cyprus ("*Southern Conveyor Project*"). Although significant efforts have been made to detect possible water leakages, as shown above, the detection of the water pipe itself it still problematic . This is because such water pipes networks are commonly mapped in a digital form (e.g. GIS environment). However, in most cases the digital location of the water pipe does not fully correspond with the real world, since many obstacles during the construction can be arise and therefore the route of the proposed pipe can change.

In order to explore further the capabilities of remote sensing –beyond the detection of water leakages- the authors have applied several algorithms for the detection of buried water pipes. The detection of buried features is well established procedure in archaeological research since buried anthropogenic remains can be found using remote sensing techniques (Agapiou *et al.*, 2010, 2012b; Sarris *et al.*, 2013). Indeed, soil marks or crop marks related with water pipes can be used, in a similar approach, for mapping the real footprint of a pipe network.

2. Study areas

In this section, three different case studies are presented. In the first case study, a part of the "*Southern Conveyor Project*" is described; following, two case studies for the "*Lakatameia* " and the "*Choirokoitia -Frenaros*" water pipes are presented. In the first case study, the authors have focused to the detection of the actual footprint of the pipe while in the next two case studies, remote sensing techniques have been evaluated for the detection of water leakages. The "*Lakatameia*"is a pipeline which is currently not in use while the "*Choirokoitia -Frenaros*" is a major pipeline of Cyprus where three major leakages have been recorded between 2007 to 2010.

2.1. "Southern Conveyor Project"

Water resources development in Cyprus initially focused on groundwater and, until 1970, groundwater was the main source of water supply for both drinking and irrigation purposes. As a result, almost all aquifers were seriously depleted because of over pumping. In addition, seawater intrusion was observed in most of the coastal aquifers. The increase of population as well as the increase in the tourist and industrial activities have led to an increase in the demand for water and have created an acute shortage of potable water.

Under these conditions, the implementation of the *"Southern Conveyor Project"* was a necessity and a basic prerequisite for the further agricultural and economic development of the island. The *"Southern Conveyor Project"* is the largest water development project ever undertaken by the Government of Cyprus. The basic objective of the project is to collect and store surplus water flowing to the sea and convey it to areas for both domestic water supply and irrigation. Essentially, the project aims to support the agricultural development of the coastal region between Limassol and Famagusta, as well as to meet the domestic water demand of Limassol, Larnaca, Famagusta, Nicosia, and a number of villages. In addition it supports the tourist and industrial demand of the southern, eastern and central areas of the island. The project is able to supply 33 million cubic metres of water for the irrigation of 13 926 hectares and another 33 million cubic metres of water for domestic purposes (Cyprus Water Development Department, 2000). In this case study, a part in the SE of Cyprus was examined (Figure 1).

2.2. "Lakatameia" pipeline

An existing pipeline in the area of Lakatameia (central Cyprus) was selected to be used for the pilot study (Figure 2). The existing pipeline, with a length of less than 5 km, has

Figure 1. Map of the SE of Cyprus showing parts of "Southern Conveyor Project" (blue line) (© Google Earth)

been system- atically reported as problematic due to several leakages and is therefore no longer in use by local authorities. The waterpipe runs through both urban and rural areas (see Figure 2). A section of the pipeline with a length of over 2km and located in a rural area, has been used to apply the different remote sensing techniques for the detection of leakages. Since the existing waterpipe is not currently used, it was necessary to fill the pipe with water periodically in order to observe the effectiveness of such remote sensing techniques.

The water pipe is made of UPVC and has 315mm diameter. It is between 1.80m and 2.00m below the ground surface and runs along the *Pediaos* river for a large part of its length. It is not currently being used due to water leakages occurring throughout almost the entire length of the pipeline. Information regarding the specific dates of the leakages is not available from local authorities.

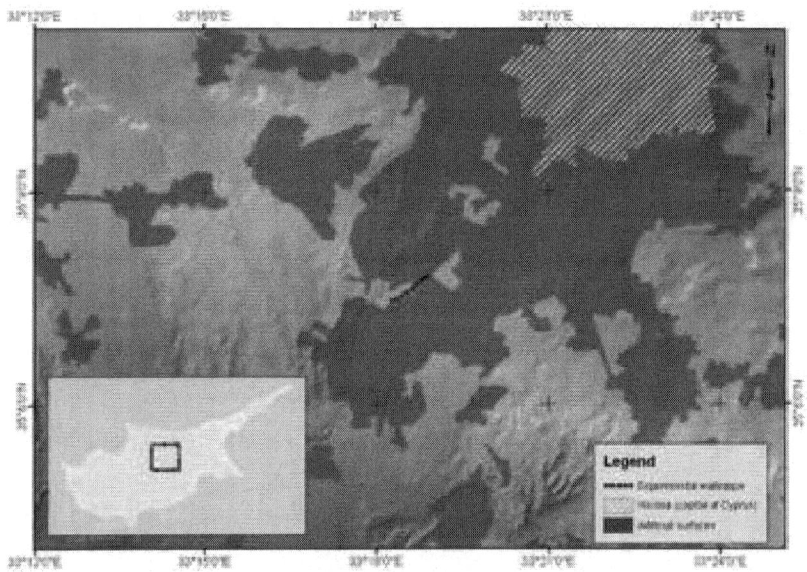

Figure 2. The "Lakatameia" waterpipe (dash line) used as the pilot study area.

2.3. "Frenaros — Choirokoitia" water pipe

The next area of interest is a major rural pipeline in Cyprus, which runs from the Choirokoitia area to the Frenaros area (Figure 3). The existing pipeline, which passes through the central and central-east part of Cyprus, has a length of over 65 km. The pipeline is located 1-3 meters below ground surface. Various geological formations, including calcaric cambisols, calcaric rego- sols, and epipetric calcisols exist in the area. elevation of the pipeline (ground surface) varies between 10 m and 200 m above sea level (Figure 4). In addition, the waterpipe passes through different types of land cover, as recorded from the CORINE 2000 land use map (Figure 5).

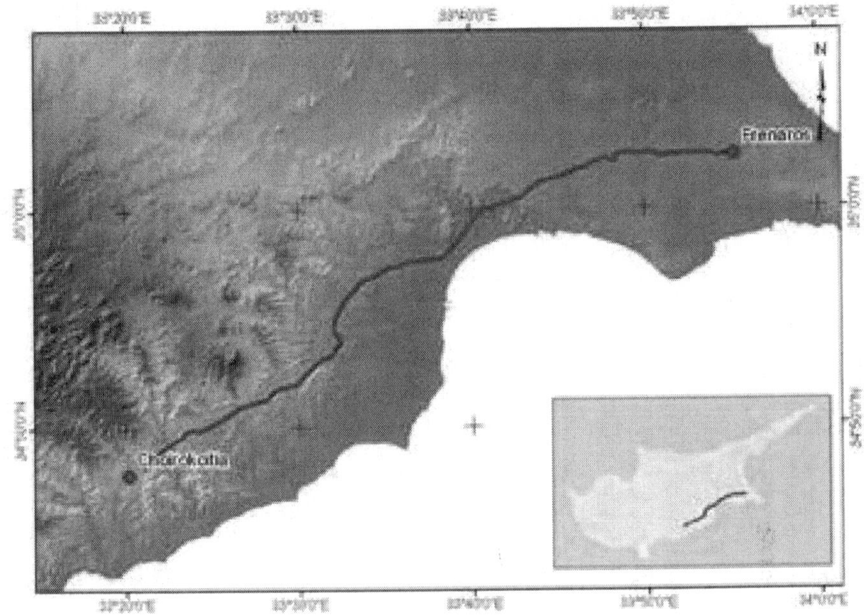

Figure 3. The "Frenaros - Choirokoitia" water pipe (solid line) used as the case study area.

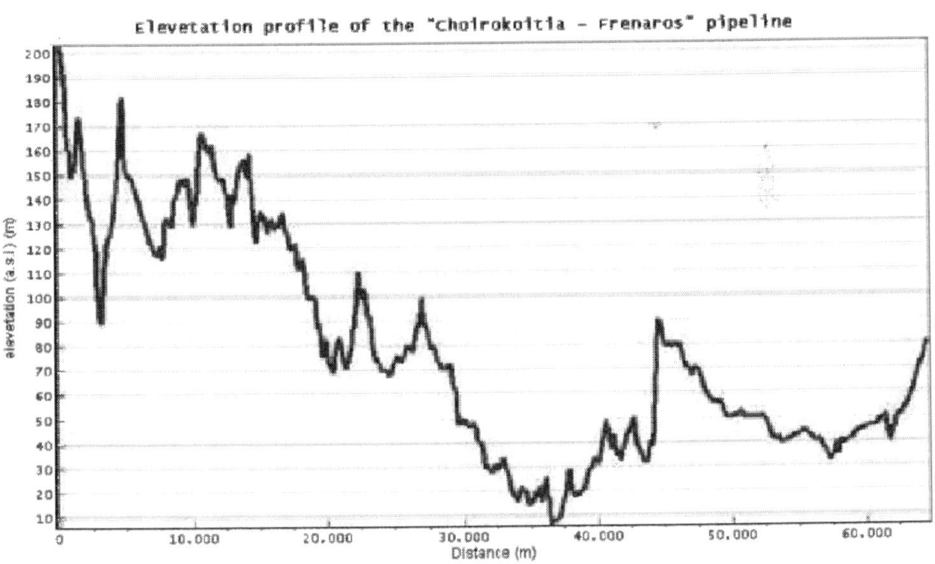

Figure 4. The elevation profile of the "Frenaros - Choirokoitia " waterpipe.

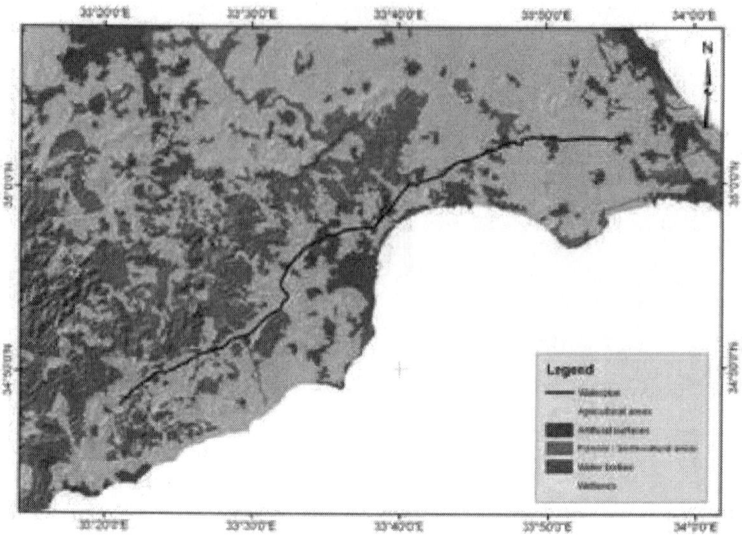

Figure 5. CORINE 2000 land use (Level 1) in the area of interest ("Choirokoitia-Frenaros" waterpipe)

During the period 2007 to 2010, three major leakages were observed along different sections of the pipe (Figure 6). The locations of these leakages were not detected until 2 months after the leakage occurred due to the difficulty of the local authorities in identifying the problematic areas. The leakages occurred during 2007; 2008 and 2010; further details for these events are presented in Table 1.

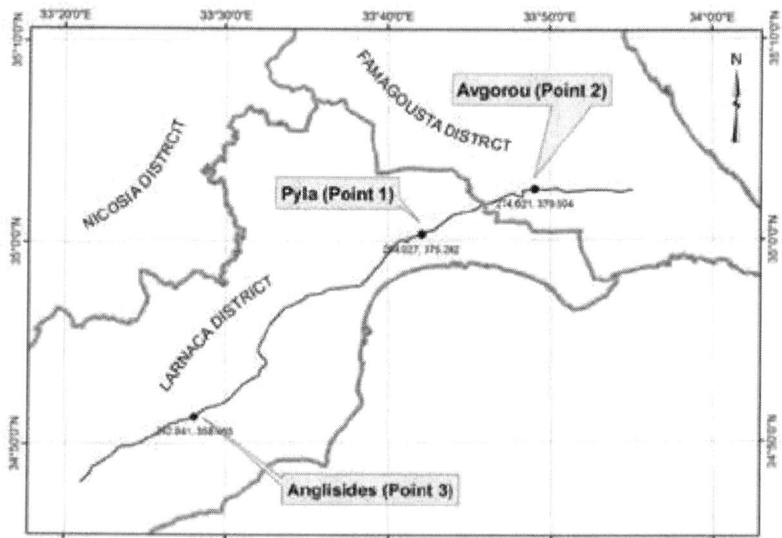

Figure 6. The "Frenaros - Choirokoitia " waterpipe (in blue). Points 1-3 indicate the three areas were water leakages have been reported.

Table 1. The leakages of the Frenaros – Choirokoitia water pipeline

Point	Position	Name	Date of pipe fixing
Point 1	Km 43.265*	Pyla Area	20-07-2007
Point 2	Km 55.346*	Avgorou area	18-02-2010
Point 3	Km 12.769*	Anglisides area	17-09-2008

* *Km positions along the pipeline, starting point Choirokoitia*

3. Methodology

The detection of the footpirnt of the *"Southern Conveyor Project"* was made based on interpre- tation techniques. The interpretation was conducted using free data from Google Earth database and using high resolution satellite images. Several histogram enhancement techni- ques were applied along with filters in order to improve the interpretation. As well, Principal Component Analysis (PCA) and classification techniques were also conducted.

In order to explore the capabilities of remote sensing for the detection of water leakages, two different methodologies were followed. For the *"Lakatameia"*waterpipe pilot study, ground spectroradiometric measurements were taken using a handheld spectroradiometer. A leakage event was created by filling several sections of the pipeline with water so that ground spectral signatures could be taken before and after the leakage. Spectroradiometric data were also recorded from different heights using a low altitude system. In this way, spectral signatures were able to simulate variation in spatial resolution (pixel size) before any other further application.

For the "*Frenaros - Choirokoitia* " water pipe case study, three major leakages have been recorded (see Table 1). Several Landsat 7 ETM+ medium resolution images, showing each leakage before and after the day the leakage was repaired, were used. A geometric and radiometric calibration of the images was performed, following by a multi-temporal analysis of all dataset based on either false composites or vegetation indices.

4. Resources

In this section, the resources and processing used for each case study are presented. The resources are grouped into three main categories: (a) high resolution satellite data used for the *"Southern Conveyor Project"* area; (b) spectroradiometric ground data used for the "*Lakata- meia*" pipeline and (c) medium resolution satellite data used for the "*Choirokoitia- Fre- naros*" pipeline.

4.1. High resolution satellite data

IKONOS high resolution satellite images were used for the detection of the buried water pipe. The IKONOS sensor, launched in 1999, was the first high-resolution satellite imagery with a spatial resolution of less than 4m. In addition, free RGB satellite images from the Google Earth databasewereexploredandanalyzed(23-10-2003;13-06-2004; 29-05-2008;30-05-2009)(Figure7).

Figure 7. IKONOS satellite image used for the detection of the buried water pipe (left) and free Google Earth images of the area (right).

4.2. Spectroradiometric data

Spectroradiometric hyperspectral measurements were carried out using the GER 1500 field spectroradiometer (Figure 8a). The GER 1500 spectroradiometer records electromagnetic radiation between 350 nm to 1050 nm (visible and near infrared part of the spectrum) A calibrated Spectralon panel, with ≈100% reflectance, was also used simultaneously to measure the incoming solar radiation. The spectralon panel measurement was used as a reference, while the measurement over the crops as a target. Therefore, reflectance for each measurement can be calculated using the following equation (1):

Reflectance = (Target Radiance / Panel Radiance) x Calibration of the panel (1)

In order to avoid any errors due to changes in the prevailing atmospheric conditions (Milton et al. 2009), the measurements over the panel and the target were taken within minutes of each other. The coordinates of the measurements were mapped using a Global Navigation Satellite Systems (GNSS) (Figure 8b).

In addition, spectroradiometric measurements were taken from a low altitude system (Figure 9). The spectroradiometer was attached to the air balloon and raised over the pilot study area. Measurements were taken at several heights in the pilot study area and also in the surrounding area in order to compare their spectral signature profiles. As the airborne system was raised, the pixel size in the ground increased. Table 2 presents some characteristic heights where the pixel size corresponds to known satellite sensors.

(a) (b)

Figure 8. (a): The GER 1500 spectroradiometer used for the collection of ground measurements and (b): the GNSS used for mapping the pipeline

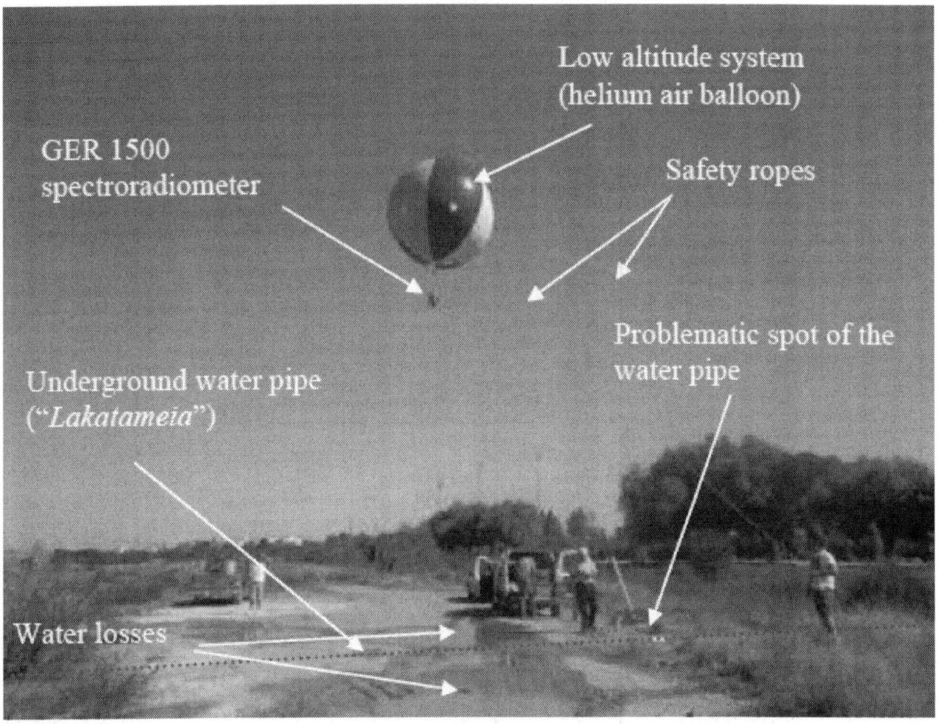

Figure 9. The low altitude system deployed over the leakage in the "Lakatameia" waterpipe.

Hyperspectral measurements recorded from the GER 1500 instrument needed to be recalcu- lated according to the characteristics of a specific multispectral satellite sensor. The authors modified these data to mimic Landsat 7 ETM+ satellite imagery based on Relative Spectral Response (RSR) filters since such data are freely distributed from the USGS. This data were used for the second case study ("Frenaros - Choirokoitia" waterpipe). RSR filters describe the instrument relative sensitivity to radiance in various parts of the electromagnetic spectrum (Wu et al. 2010). These spectral responses have a value of 0 to 1 and have no units since they are relative to the peak response (Figure 10). Bandpass filters are used in the same way in spectroradiometers in order to transmit a certain wavelength band and block others. The reflectance from the spectroradiometer was calculated based on the wavelength of each sensor and the RSR filter as follows:

Table 2. Pixel size from different heights using the low altitude system. The right column presents active satellite sensors with similar spatial resolution. Two lens with different field of view (FOV) have been be used in the GER 1500 spectroradiometer

Height from the ground	4 FOV (ground pixel - m)	8° FOV (ground pixel -m)	Satellite sensor
5	0.3	0.7	GeoEye (pan); WorldView-1
10	0.7	1.4	IKONOS (pan)

15	1.0	2.1	
20	1.4	2.8	ALOS (pan)
25	1.7	3.5	
50	3.5	7.0	IKONOS (multi)
75	5.2	10.5	ALOS (multi)
100	7.0	14.0	
150	10.5	21.0	Landsat (pan)
200	14.0	28.0	IKONOS (multi)

$$Rband = \Sigma(Ri * RSRi)/\Sigma RSRi \quad (2)$$

Where:

Rband = reflectance at a range of wavelength (e.g. Band 1)

Ri = reflectance at a specific wavelength (e.g R 450 nm)

RSRi = Relative Response value at the specific wavelength

4.3. Medium resolution satellite data

Twelve medium resolution Landsat 7 ETM+ satellite images were used, dated before and after the local authorities fixed the leaks on the "*Frenaros-Choirokoitia*" pipeline (Figure 11; Table 3). ERDAS Imagine v. 10 software was used for the pre- and post-processing of satellite imagery. Pre-processing included geometric and atmospheric correction correction of the satellite imagery. Geometric correction of the satellite images was conducted using ground control points (GCPs), which included environmental features and ground coordinates. The Darkest Pixel (DP) atmospheric correction method was used, which is the most widely applied method of atmospheric correction that provides reasonable correction (Hadjimitsis et al., 2004; Hadjimitsis et al., 2009).

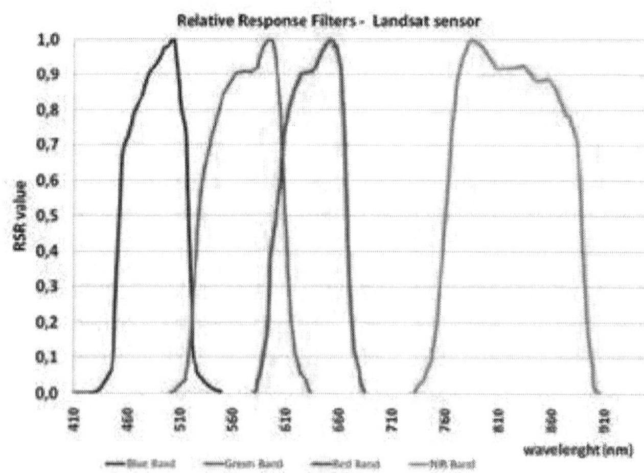

Figure 10. Relative Response filters for Bands 1-4 of Landat TM sensor (Alexakis et al. 2012)

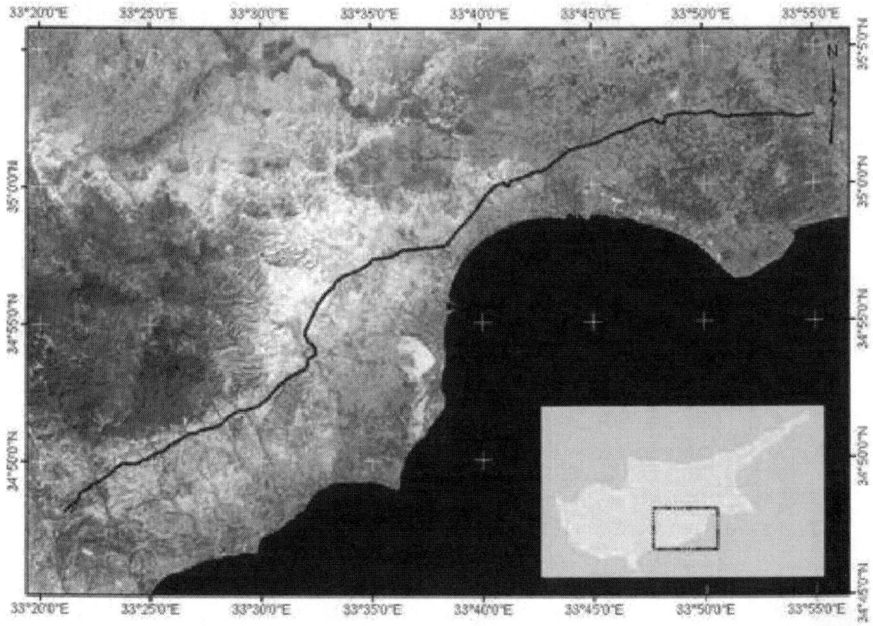

Figure 11. Landsat 7 ETM+ satellite image (28/07/2008) over the "Choirokoitia-Frenaros" water pipe.

After the necessary pre-processing steps, several vegetation indices were evaluated. False colour composites were also applied in order to detect the water leakages from the entire dataset. The evaluation was made not only in the three areas of interest (leakage problem) but also along the entire length of the water pipe. The results were mapped and statistical analysis was performed.

Table 3. Satellite images used for this study

no	Satellite	Overpass	no	Satellite	Overpass
1	Landsat ETM+	07/05/2007	7	Landsat ETM+	14/09/2008
2	Landsat ETM+	23/05/2007	8	Landsat ETM+	30/09/2008
3	Landsat ETM+	27/08/2007	9	Landsat ETM+	16/10/2008
4	Landsat ETM+	28/07/2008	10	Landsat ETM+	22/12/2009
5	Landsat ETM+	13/08/2008	11	Landsat ETM+	07/01/2010
6	Landsat ETM+	29/08/2008	12	Landsat ETM+	13/04/2010

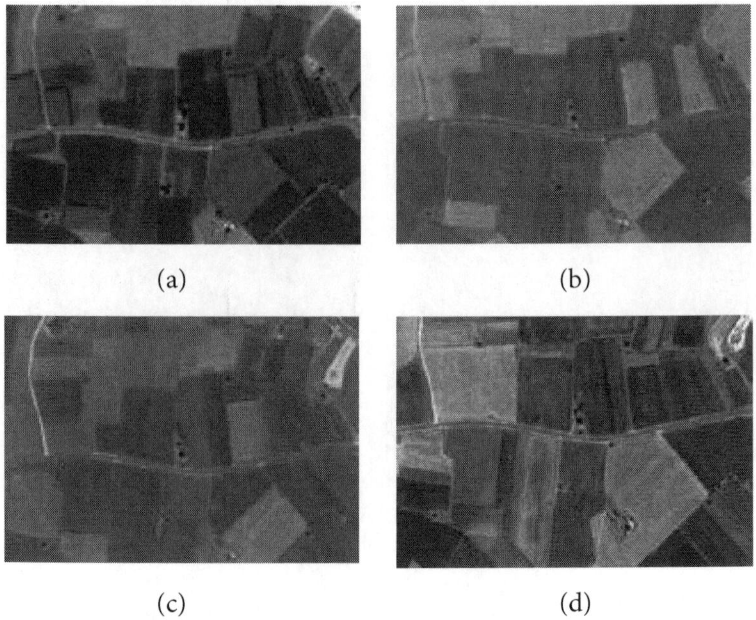

Figure 12. Google Earth satellite image used for the detection of the buried water pipe during different periods: (a): 23-10-2003; (b): 13-06-2004; (c): 29-05-2008; and (d): 30-05-2009.

5. Results

5.1 "Southern Conveyor Project" pipeline

The detection of the buried water pipe was initially performed using the multi-temporal Google Earth images (Figure 12). As shown, the success rate for the detection of the

Figure 13. The footprint of the water pipe as a result of the tree pattern.

water pipe can vary depending on the period of observation. The interpretation could be performed much easier in areas with no coverage (bare soil) while in cultivated areas the interpretation was a difficult task. In addition, images taken just after rainfall or after watering crops, tend to provide better results since soil marks could be easily spotted.

Moreover, the tree pattern could reveal the footprint of the water pipe (see Figure 13). This pattern can be used for the detection of buried water pipes or can be used for monitoring possible problems resulting from tree roots.

The IKONOS image used for this case study was able to maximize the visible footprint of the water pipe. Indeed, using the VNIR part of the spectrum and false colour composites (Figure 14) made possible the detection of both soil and crop marks. The IKONOS multispectral image was able to detect other parts of the water pipe network of the area, as shown in Figure 14 (right arrow). Spatial filter and PCA analysis applied to the image data (Figure 15) were able further to enhance the interpretation.

In an attempt to evaluate if an automatic detection of such crop marks could be performed (e.g. classification), spectral profiles were examined. Spectral signatures from the image were evaluated as shown in Figure 15, which features areas of crop marks and of healthy vegetation. Scatter plots from these two areas (Figure 16) indicate that a spectral difference exists between these areas, especially in the VNIR part of the spectrum.

Figure 14. IKONOS VNIR-R-G pseudo colour composite

Figure 15. IKONOS 3 x 3 high pass filter (left) and PCA analysis (right)

5.2. "*Lakatameia*" pipeline

The results found that water leakages could be monitored using remote sensing techniques. As shown in Figure 17, the spectral signatures of dry and wet soil is easily recognized in the visible range of the spectrum (400 -700 nm) and in the very near infrared range (750-900nm). Wet soil tends to give 20-25% lower reflectance values compare to the dry soil. This difference is also maximized in the very near infrared range of the spectrum. Similarly, Figure 18 indicates spectral signature profiles of several targets before (dry) and after (wet) a leakage event. Similar findings also applied to vegetation. Dry grass tends to give approximately 5% reflectance in the green part of the spectrum (520-600nm) and 25% in the very near infrared (750-900nm) in contrast to 12% and 35% respectively for the wet grass.

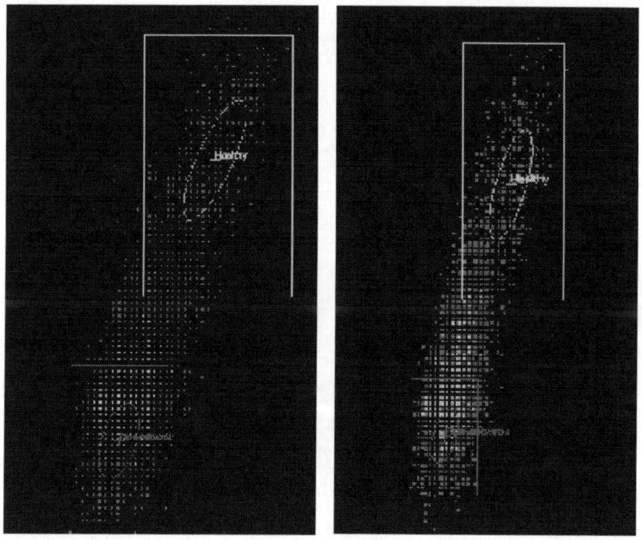

Figure 16. Scatter plots from crop marks (red square) and healthy vegetation (yellow square) for Bands 1-3 and Bands 1-4 combinations (left and right respectively).

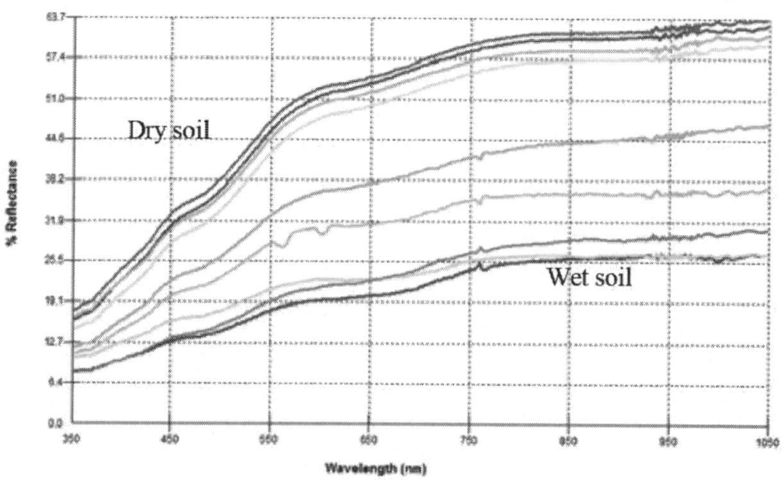

Figure 17. Ground spectral signatures over dry and wet soil in the 'Lakatameia' pipeline

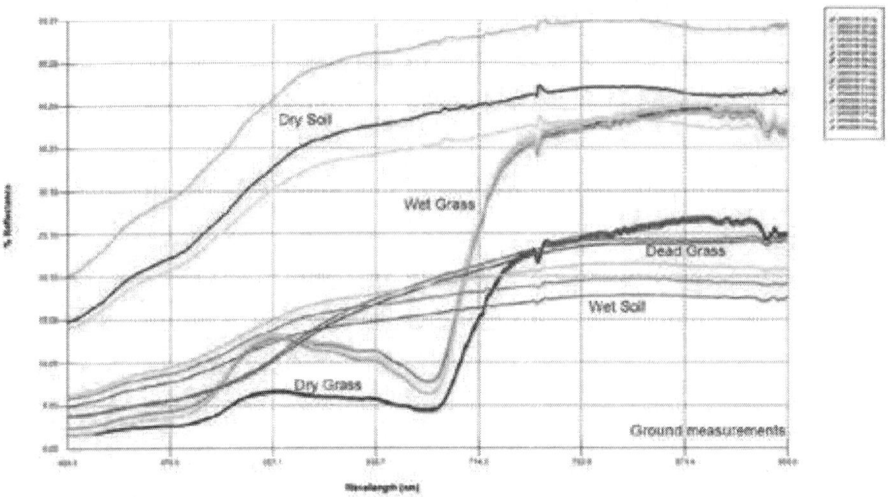

Figure 18. Ground spectral signatures of different targets in the 'Lakatameia' pipeline

Figures 19 and 20 present the spectral signatures over the same areas from different heights, using the low altitude system. Reflectance initially increases as the system is raised above ground level (until 10 meters) while a small decrease of the reflectance is observed afterwards (16 meters) which can be associated with the larger area covered from the spectroradiometer. However it should be noted that these differences (~5%) are similar to the total relative uncertainties of calibration for satellite sensors (within 5%) (Trishchenko et al. 2002).

The above results are well supported in the literature. Nocita et al. (2011), Ouillon et al. (2002), Dobos (2003), Kaleita et al. (2005) and Garcia-Rodriguez (2011) found that moisture affects the reflectance value of soil. There is a notable decrease in reflectance with increasing moisture in the ground (Bowers and Hanks, 1965; Baumgardner et al., 1985; Twomey et al., 1986; Ishida et al., 1991; Whiting et al., 2000; Bogrekci and Lee, 2005; Lesaignoux et al. 2007). However, the rate of decrease in relative reflectance becomes more moderate with increasing ground moisture, since at very high moisture contents, the soil is already quite dark and further moisture added to the soil has less of an effect on the reflectance (Kaleita et al., 2005). Moisture dominates the spectral reflectance of soils in the 340-2500 nm wavelengths (Somers et al., 2010; Bogrekci and Lee, 2005). Moisture affects the reflection of shortwave radiation from ground surfaces in the visible and near-infrared - VNIR (400-1100nm) and shortwave infrared - SWIR (1100-2500nm) regions of the spectrum (Bowers and Hanks, 1965; Skidmore et al., 1975). It is notable that, although precipitation affects the reflectance value for each target, it does not change the typical spectral signature between wet and dry conditions (Philpot, 2010).

The results indicate that the detection of a leakage event is possible using remote sensing techniques. Indeed, the use of the very near infrared range of the spectrum can be used on areas with bare soil or with vegetation. The findings from this pipeline were therefore compared with data from actual cases studies of water leakage in the '*Freanaros-Choirokoitia*' pipeline.

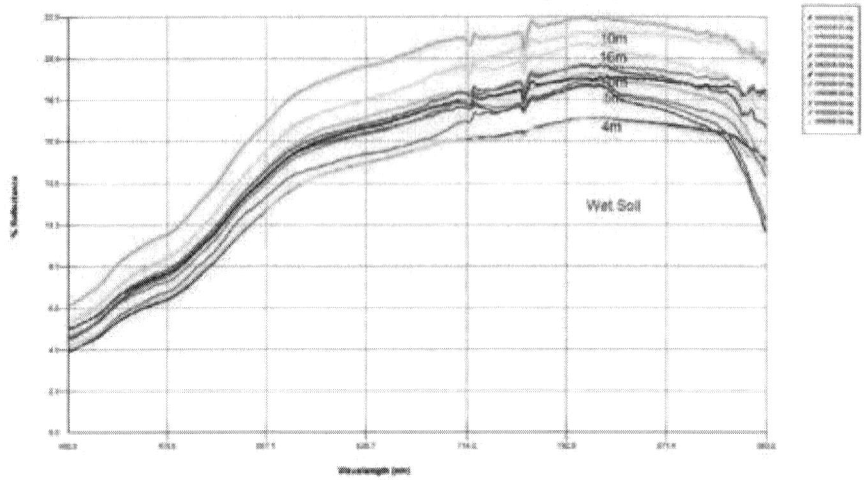

Figure 19. Spectral signatures of wet soil in the 'Lakatameia' pipeline at different heights using the low altitude system

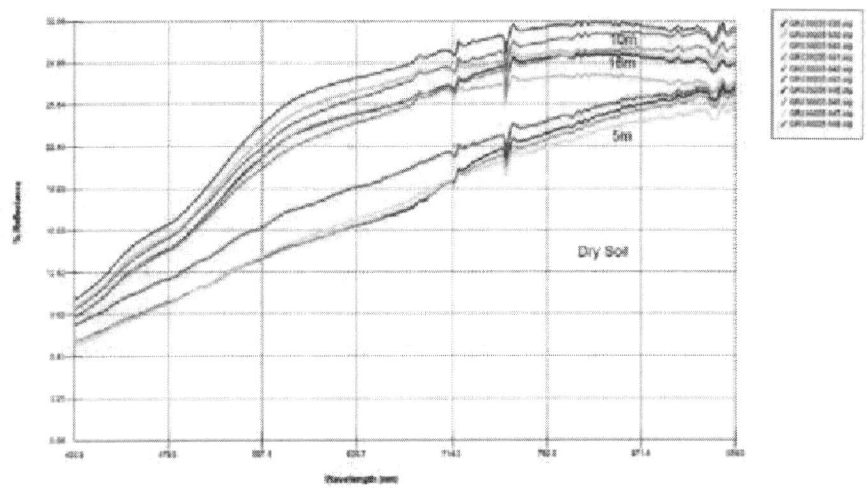

Figure 20. Spectral signatures of dry soil in the 'Lakatameia' pipeline at different heights using the low altitude system

5.3. "Frenaros — Choirokoitia " water pipe

Based on the findings of the *"Lakatameia"* water pipe, satellite images where used for the detection of known water leakages using archive satellite images. In order to examine the capabilities of satellite remote sensing images for the detection of water leakages, several algorithms and analyses were carried out. At first, reflectance values of all datasets (see Table 3) were calculated based on the metadata file using equations 3 and 4. Following this, several vegetation indices were calculated. In addition, different false colour composites were produced to assess the ability of the system to detect the known leakages from the satellite images.

For Point 1 at *Pyla* area, leakage detection was difficult using medium resolution images. Monitoring of the pipeline using the red and the near infrared part of the spectrum for Point 1 did not reveal any significant changes of reflectance due to the water leakage. Similarly, vegetation indices (NDVI) did not show any differences for Point 3 (*Anglisides* area).

However, for Point 2, Landsat 7 ETM+, promising results were found. As shown in Figure 21, the Landsat satellite image dated January 7, 2010, tends to give higher vegetation index values, prior to the water leakage being repaired on February 18, 2010. However, the above hypothesis is applicable to other areas of the water pipe as well. The above results have shown the limitations of using medium resolution satellite images for the detection of water leakages, especially when these are rare and small.

Point 2: "Frenaros-Choirokoitia" pipeline

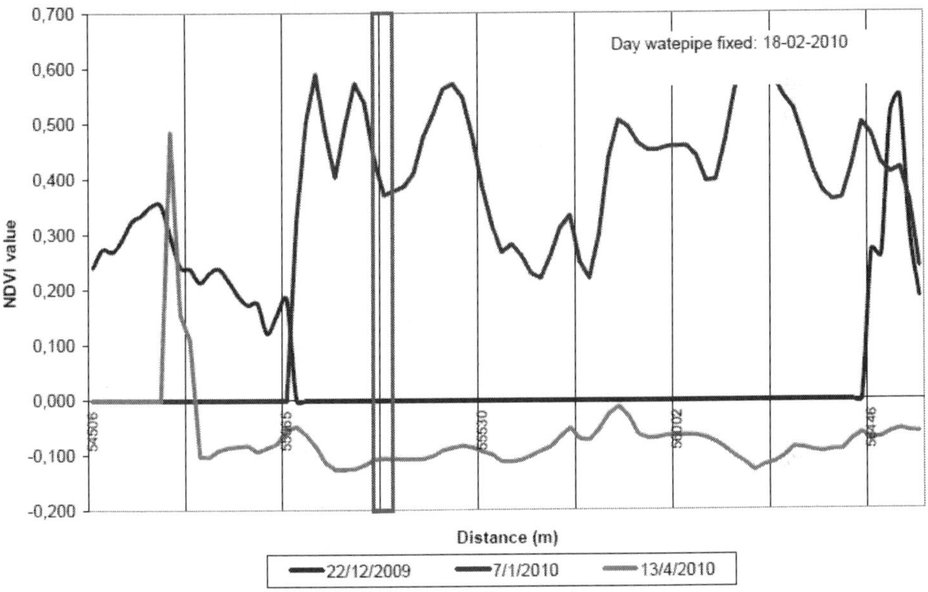

Figure 21. NDVI values using Landsat 7 ETM+ images used over Point 2. The red square highlights the area where the leakage was observed.

In an effort to explore further the information extracted using satellite data the three pilot areas were examined separately. Three vegetation indices, the

Normalized Difference Vegetation Index (NDVI);

Soil Adjusted. Vegetation Index (SAVI) and the

Ratio Vegetation Index (RVI) were calculated based on the formulas shown in equations 5, 6 and 7.

$$\left(p_{NIR} - p_{red}\right) / \left(p_{NIR} + p_{red}\right) \tag{3}$$

$$(1{+}0.5)\,(p_{NIR} - p_{rb})/(\,p_{NIR} + p_{red} {+}0.5) \tag{4}$$

$$p_{red}/\,p_{NIR} \tag{5}$$

Where:

p_{NIR} is the near infrared reflectance p_{red} is the red reflectance

Figure 22 presents the NDVI development during the examined 12 dates of satellite overpasses (see Table 3). Figure 23 presents the SAVI development during the examined 12 dates of satellite overpasses.

Point 1,2,3: "Frenaros-Choirokoitia" pipeline

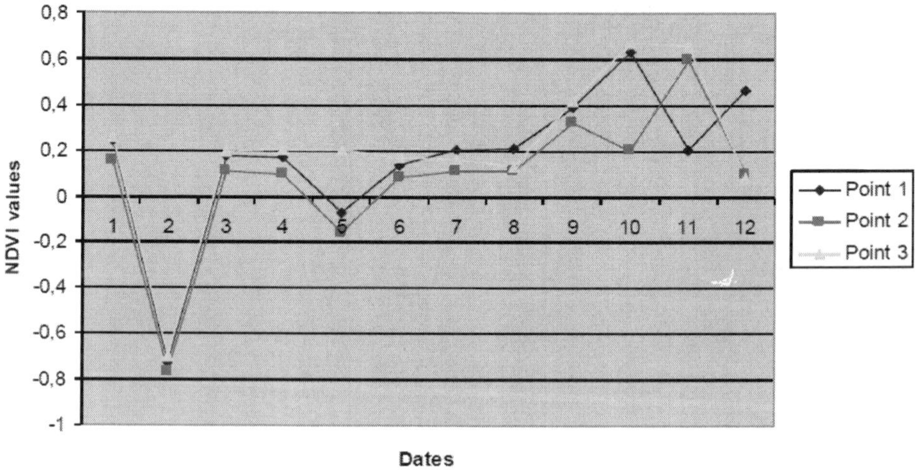

Figure 22. NDVI refl. (calculated with Reflectance values) development during the examined 12 dates (Landsat im- ages) in Points 1, 2 and 3

Points 1,2,3: "Frenaros-Choirokoitia" pipeline

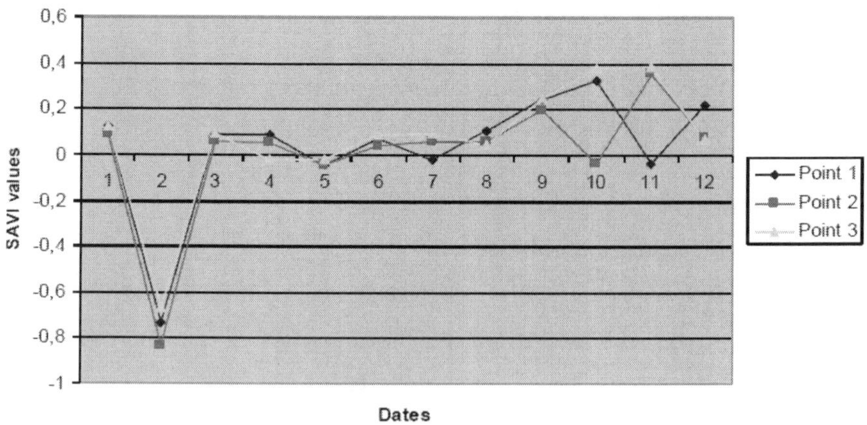

Figure 23. SAVI refl. (calculated with Reflectance values) development during the examined 12 dates (Landsat im- ages) in Points 1, 2 and 3

Based on the graphs of Figure 22, the NDVI values present the following pattern: during May 2007, in all 3 points of the known water leakage, NDVI decreases significantly with values close to -0.8, when almost in all cases NDVI is above zero with similar values. After September 2008, the NDVI values increase until April 2010 when they decline again. Such results indicate that the vegetation of the area around the study points reflects soil moisture resulting from rainfall as it can differentiate according to season. Detailed examination of each point related to the pipeline repair indicates that NDVI in Points 1 and 2, water leakage ceased just after the 2nd and the 11th date in correspondence: (a) Point 1: -0,72 and 0,17 for days 2 and 3 and (b) Point 2: 0,60 and 0,09 for days 11 and 12 respectively.

For Point 1, there is a significant change of NDVI value before and after the repair date of the pipeline. In Point 2, the NDVI value decreased significantly (from 0, 60 to 0, 09) following the repair of the pipeline.

However, in Point 3 there is no significant change of the NDVI value before and after the repair date of the pipeline. Although there is a slight decrease in NDVI values immediately following the repair, there is a significant increase within 2 weeks: Point 3: 0,16; 0,13 and 0,42 for days 7 -9 respectively.

The results indicate that only at Point 2 is there a significant decline of NDVI values as a result of lack of soil moisture around the pipe. Another factor can be that due to the temporal difference between the two measurements, of 7 January 2010 and 13 April 2010, respectively, as lack of rainfall may have resulted in moisture evaporation. The same conclusion is reached with SAVI data (Figure 23). The value of SAVI in Point 2 was 0,35 in January 2010 and declined to 0,07 just after the pipeline repair.

Figure 24 presents RVI data which were calculated using equation 7. The RVI index indicates the effect of soil moisture around Point 2. The RVI value in Point 2, in January 2010 was 4,02 and after the pipeline repair, it decreased to 1,21. It seems that the

Points 1,2,3: "Frenaros-Choirokoitia" pipeline

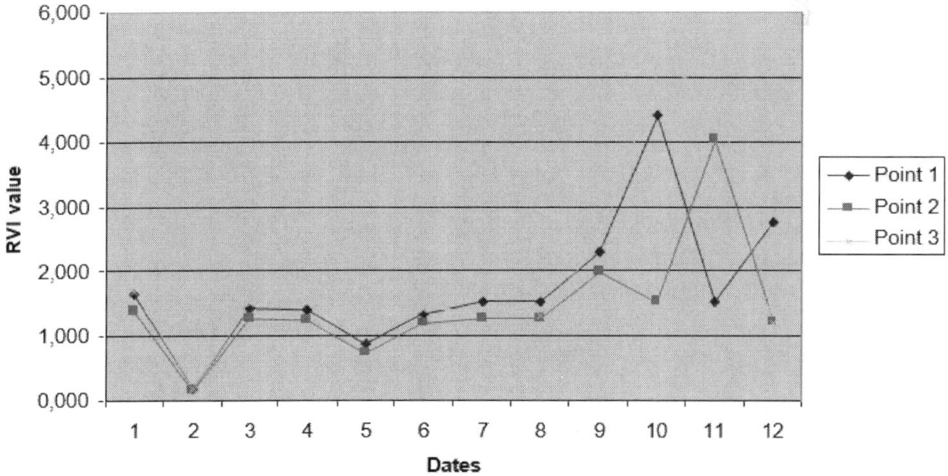

Figure 24. RVI refl. (calculated with Reflectance values) development during the examined 12 dates (Landsat images) in Points 1, 2 and 3

vegetation developed on the soil around Point 2, and subsequently dried after the repair of the water pipe and the evaporation of the soil water.

In addition, meteorological data provided from the Meteorological Service of Cyprus, indicate that significant rainfall was recorded on 25, 26 and 27 February 2010, after the pipe line repair date of Point 2 (18 February, 2010). During March and April of 2010, only 1.0 and 2.1 mm of rain were recorded for the same location. Such information provides additional validation that the main factor affecting the NDVI, SAVI and RVI values is the presence or absence of vegetation as a result of soil moisture before and after the pipeline leakage repair. Regarding Point 3, in the Anglisides area, September precipitation data did not affect the pipe leakage since no significant rainfall was recorded before and after the pipeline repair (17 September, 2008).

6. Discussion and remarks

Remote sensing techniques have been found to be effective both for the detection of the water pipes and for the detection of water leakages. The preliminary results of this study have shown that remote sensing techniques are able to detect areas of the pipeline with water leakages. Ground spectroradiometric data along with the low altitude spectroradiometer system indicate significant differences in the reflectance values in areas where leakage is observed. In addition, crop and soil marks can be used for mapping the actual footprint of the water pipe.

Although the use of medium resolution satellite images for monitoring extensive pipelines may be problematic, such as in Points 1 and 3 in the "*Franaros - Choirokoitia*" pipeline, this may be due to the spatial resolution of the specific satellite images. However, promising results have been also reported (i.e. Point 2 in the "*Franaros - Choirokoitia*" pipeline), where a major leakage was observed.

In addition, remote sensing techniques can be used on a systematic basis to monitor specific problematic areas of a water network by using time-series satellite images. Future research will investigate additional ground based geophysical methods to provide a competent system for monitoring existing water pipe networks, such as electrical resistance tomography and ground penetrating radar. The resulting data can be integrated into a Geographical Informa- tion System which can be used by local authorities.

References

1. Agapiou, A, Hadjimitsis, D. G, & Alexakis, D. D. (2012a). Performance of vegetation indices for supporting ground and satellite remote sensing archaeological investigations. Remote Sensing. doi:10.3390/rs40x000x, 4
2. Agapiou, A, Hadjimitsis, D. G, Alexakis, D, & Sarris, A. (2012b). Observatory validation of Neolithic tells ("Magoules") in the Thessalian plain, central Greece, using hy- perspectral spectro-radiometric data, Journal of Archaeological Science, doi.org/ 10.1016/j.jas.2012.01.001., 39(5), 1499-1512.
3. Agapiou, A, Hadjimitsis, G. D, Sarris, A, Themistocleous, K, & Papadavid, G. (2010). Hyperspectral ground truth data for the detection of buried architectural remains, Lecture Notes in Computer Science, , 6436, 318-331.

4. Ahadi, M, & Bakhtiar, S. M. (2010). Leak detection in water-filled plastic pipes through the application of tuned wavelet transforms to Acoustic Emission signals. Applied Acoustics, , 71, 634-639.

5. Alexakis, D, Agapiou, A, Hadjimitsis, D. G, & Sarris, A. (2012). Remote sensing applications in archaeology. Remote Sensing / Book 2 (979-9-53307-231-8Book edited by: Boris Escalante.

6. Bannari, A, Morin, D, Huette, A. R, & Bonn, F. (1995). A review of vegetation indices. Remote Sensing Reviews, , 13, 95-120.

7. Baumgardner, M. F, Silva, L. F, Biehl, L. L, & Stoner, E. R. (1985). Reflectance properties of soils. Advanced Agronomy, , 38, 1-44.

8. Bogrekci, I, & Lee, W. S. (2006). Effects of Soil Moisture Content on Absorbance Spec- tra of Sandy Soils in Sensing Phosphorus Concentrations Using UV-VIS-NIR Spectro- scopy. Transactions of the American Society of Agricultural Engineers, , 49, 1175-1180.

9. Bowers, S. A, & Hanks, R. J. (1965). Reflection of radiant energy from soils. Soil Science, , 2, 130-138.

10. Burn, L. S, Davis, P, Desilva, D, Marksjo, B, Tucker, S. N, & Geehman, C. J. (2001). The Role of Planning Models in Pipeline Rehabilitation. Plastic Pipes XI,th September 2001. Munich, Germany., 3-6.

11. Dobos, E. (2003). Albedo, Encyclopedia of Soil Science, DOI:E-ESS 120014334., 1-3.

12. Eyuboglu, S, & Mahdi, H. dan Al-Shukri, H. ((2003). Detection of Water Leaks using Ground Penetrating Radar. The 3rd International Conference on Applied Geophy- sics. December Orlando, Florida: Environmental and Engineering Geophysical Soci- ety., 8-12.

13. Faidrullah, S. N. (2007). Normalized Different Vegetation Index for Water Pipeline Leakage Detection. In: The 28th Asian Conference on Remote Sensing November 2007, PWTC, Kuala Lumpur, Malaysia., 2007, 12-16.

14. García RodríguezJ. N. ((2011). Changes in Spectral Slope due to the Effect of Grain Size and Moisture in Beach Sand of Western Puerto Rico. Accessed 21 October, 2011: http://gers.uprm.edu/pdfs/topico_johanna2.pdf.

15. Hadjimitsis, D. G, Clayton, C. R. I, & Retalis, A. (2009). The use of selected pseudo-invariant targets for the application of atmospheric correction in multi-temporal studies using satellite remotely sensed imagery, International Journal of Applied Earth Observation and Geoinformation, DOI:j.jag.2009.01.00., 11, 192-200.

16. Hadjimitsis, D. G, Clayton, C. R. I, & Hope, V. S. (2004). An assessment of the effectiveness of atmospheric correction algorithms through the remote sensing of some reservoirs. International Journal of Remote Sensing, DOI:, 25, 3651-3674.

17. Hadjimitsis, D. G, Themistocleous, K, & Achilleos, C. (2009). Integrated use of GIS, GPS and Sensor Technology for managing water losses in the water distribution network of the Paphos Municipality in Cyprus, STATGIS 2009, Milos-Greece, June, 2009., 17-19.

18. Heathcote, M, & Nicholas, D. (1998). Life assessment of large cast iron watermains. Urban Water Research Association of Australia (UWRAA), Research Report (146)

19. Huang, Y, Fipps, G, Maas, J. S, & Fletcher, S. R. (2010). Airborne remote sensing for detection of irrigation canal leakage. Irrigation and Drainage, , 59, 524-553.

20. Hunaidi, O, & Giamou, P. (1998). Ground Penetrating Radar for detection of leaks in buried plastic water distribution pipes. Seventh International Conference on Ground Penetrating Radar,Lawrence, Kansas, USA, , 27-30.

21. Ishida, T, Ando, H, & Fukuhra, M. (1991). Estimation of complex refractive index of soil particles and its dependence on soil chemical properties. Remote Sensing of En- vironment, , 38, 173-182.

22. Kaleita, A. L, Tian, L. F, & Hirschi, M. C. (2005). Relationship between soil moisture content and soil surface reflectance. Transactions of the American Society of Agricultural Engineers, , 48, 1979-1986.

23. Lesaignoux, A, Fabre, S, Briottet, X, Olioso, A, Belin, E, & Cedex, T. (2009). Influence of surface soil moisture on spectral reflectance of bare soil in the 0.m domain. Proceedings of the 6th EARSeL SIG IS workshop imaging spectroscopy., 4-15.

24. Milton, E. J, Schaepman, M. E, Anderson, K, Kneubühler, M, & Fox, N. (2009). Progress in Field Spectroscopy. Remote Sensing of Environment, , 113, 92-109.

25. Nocita, M, Stevens, A, & Van Wesenmael, B. (2011). Improving spectral techniques to determine soil organic carbon by accounting for soil moisture effects. The Second Global Workshop on Proximal Soil Sensing, Montreal.

26. Ouillon, S, Lucas, Y, & Gaggelli, J. (2002). Hyperspectral detection of sand. Presentation at the Seventh International Conference on remote sensing and coastal environments. Miami Florida, May , 20-22.

27. Philpot, W. (2010). Spectral Reflectance of Wetted Soils. Proceedings of ASD and IEEE GRS; Art, Science and Applications of Reflectance Spectroscopy Symposium, Vol. II.

28. Pickerill, J. M, & Malthus, T. J. (1998). Leak detection from rural aqueducts using air- borne remote sensing techniques. International Journal of Remote Sensing, , 19, 2427-2433.

29. Poulakis, Z, Valougeorgis, D, & Papadimitriou, C. (2003). Leakage detection in water pipe networks using a Bayesian probabilistic framework. Probabilistic Engineering Mechanics, , 18, 315-327.

30. Sarris, A, Papadopoulos, N, Agapiou, A, Salvi, C. M, Hadjimitsis, D. G, Parkinson, A. W, Yerkes, W. R, Gyucha, A, & Duffy, R. P. (2013). Integration of geophysical sur- veys, ground hyperspectral measurements, aerial and satellite imagery for archaeo- logical prospection of prehistoric sites: the case study of Vésztő-Mágor Tell, Hungary, Journal of Archaeological Science, doi:j.jas.2012.11.001., 40, 1454-1470.

31. Skidmore, E. L, Dickerson, J. D, & Shimmelpfennig, H. (1975). Evaluating surface-soil water content by measuring reflectance. Soil Science Society Annual Proceedings, , 39, 238-242.

32. Skolnik, M. I. (1990). Radar Handbook, 2nd Ed. New York: McGraw-Hill.

33. Somers, B, Tits, L, Verstraeten, W. W, & Coppin, P. (2010). Soil reflectance modeling and hyperspectral mixture analysis: towards vegetation spectra minimizing the soil background contamination. Hyperspectral Image and Signal Processing: Evolution in Remote Sensing (WHISPERS), June 2010., 1-4.

34. Twomey, S. A, Bohren, C. F, & Mergenthaler, J. L. (1986). Reflectances and albedo di- ferences between wet and dry surfaces. Applied Optics, , 25, 431-437.

35. Trishchenko, P. A, Cihlar, J, & Zhanqing, L. (2002). Effects of spectral response function on surface reflectance and NDVI measured with moderate resolution satellite sensors. Remote Sensing of Environment, , 81, 1-18.

36. Weifeng, L, Wencui, L, Suoxiang, L, Jing, Z, Ruiping, L, Qiuwen, C, Zhimin, Q, & Jiu- hui, Q. (2011). Development of systems for detection, early warning, and control of pipeline leakage in drinking water distribution: A case study. Journal of Environmental Sciences, , 23, 1816-1822.

37. Whiting, M. L, Li, L, & Ustin, S. L. (2003). Estimating surface soil moisture in simulat- ed AVIRIS spectra. Twelfth Annual Airborne Earth Science and Application Work- shop. Jet Propulsion Laboratory, California Institute of Technology, Pasadena, California, February , 25-28.

38. Wu, X, Sullivan, T. J, & Heidinger, K. A. (2010). Operational calibration of the advanced very high resolution radiometer (AVHRR) visible and near-infrared chan- nels. Canadian Journal of Remote Sesning, , 36, 602-616.

Chapter-7

Air Pollution from Space

1. Introduction

The South Eastern Mediterranean region is an atmospheric cross road where aerosols of different origins can be observed. Atmospheric pollution due to particulate matter from natural and anthropogenic sources is a continuing problem in many areas of Cyprus. Partic- ulate matter (PM) is a major component of urban air pollution and has a significant effect on human health. High quality PM monitoring with a fine spatial and temporal resolution may help decision makers to assess the efficiency of control strategies and also may be useful for informing the general public about air pollution levels and hazards. The AIRSPACE research project was established with the main aim of combining remote sensing data (mainly MODIS) with concurrent in-situ observations (sunphotometric, LIDAR and ground level PM measure- ments) for monitoring air pollution in an integrated manner. AIRSPACE aims to develop a novel methodology based on in-situ experimental observations in order to use satellite retrieval as a tool for monitoring air particulate pollution. This methodology was applied in Cyprus with an emphasis on urban areas and, to a lesser extent, industrial regions. Observa- tions from passive and active ground-based and satellite techniques for Aerosol Optical Thickness (AOT) retrieval, in combination with PM_{10} and $PM_{2.5}$ concentrations at sites near different PM sources, have been considered. Several factors, such as aerosol vertical distribu- tion, that affect the relationship between PM ground measurements and AOT, were consid- ered. Data sets from three types of sites (urban, near urban and rural) were used to develop a statistical model for the estimating PM mass concentrations using AOT measured from remote sensing techniques and meteorological parameters. Furthermore, the ground truth observa- tions collected within AIRSPACE project were used to assess qualitative and quantitative performance of a chemical model forecast of PM concentrations throughout Cyprus.

Midletton et al. (2008) reported that in Nicosia, Cyprus for every 10-$\mu g/m^3$ increase in PM_{10} daily average concentrations there was a 0.9% (95%CI: 0.6%, 1.2%) increase in all-cause and 1.2% (95%CI: -0.0%, 2.4%) increase in cardiovascular admissions. A recent study regarding dust storm events in Nicosia, Cyprus, found a 2.43% (95% CI: 0.53, 4.37) increase in daily cardiovascular mortality associated with each 10-$\mu g/m^3$ increase in PM_{10} concentrations on dust days in comparison with non-dust days (Neophytou et al., 2013).

2. Background

Air pollution in large cities is one of the major issues to be addressed by local and global communities due to its widespread presence, and deleterious impact on human life (Hadji- mitsis, 2009). As air pollution is a major environmental health risk, by reducing the levels of air pollutants, countries will reduce the incidence of disease from respiratory infections, heart disease and lung cancer (WHO, 2011). Actions by policy makers and public authorities at the national, regional and international levels are required in order to control the exposure to air pollutants (EEA, Air Quality in Europe, 2012 - Report). Transboundary domestic air pollution is of high concern among the EU member states. In 2010, about 21% of the EU urban population was exposed to concentrations of PM_{10} above the limit value established by the European Environmental Agency (EEA, Air Quality in Europe, 2012 - Report). The WHO, USEPA (U.S. Environmental Protection Agency) and EEA have established an extensive body of legislation which establishes standards and objectives for a number of air pollutants such as PM_{10} (coarse particles), $PM_{2.5}$ (fine particles) and O_3 (WHO, Fact sheet No 313, 2011; USEPA, NAAQS, 2012; EEA, AAS, 2012).

Current research focused on the study of regional and intercontinental transport of air pollutants, such as particulate matter ($PM_{10, 2.5}$), points to a need for additional data sources to monitor air pollution in multiple dimensions, both spatially and temporally. To address this issue, earth observations from satellite sensors can be a valuable tool for monitoring air pollution due to their ability to provide complete and synoptic views of large areas.

Although air quality monitoring stations have been established in major cities, there is an increased need to establish mobile stations for additional coverage, as such stations provide a means for alerting the public regarding air quality. However, measuring stations are localised and do not provide sufficient geographic coverage, since air quality is highly variable spatially. The use of earth observations to monitor air pollution in different geographical areas, espe- cially cities, has received considerable attention from researchers (see Wald et al., 1999; Grosso and Paronis, 2012; Hadjimitsis, 2009; Hadjimitsis et al., 2010; Jones and Christopher, 2007; Michaelides et al., 2011; Nisantzi et al., 2012; Retalis and Sifakis, 2010; Retalis et al., 2003; Retalis et al., 1999; Vidot et al., 2007). Several researchers (Chudnovsky et al., 2013; Gupta et al., 2006; Koelemeijer et al., 2006; van Donkelaar et al., 2010) have focused on the use of satellite sensors on air pollution studies, especially their ability for systematic monitoring and synoptic coverage. The use of sunphotometers and LIDAR systems are found to be suitable tools for assisting the air pollution monitoring studies (Ansmann et al., 2012; Amiridis et al., 2008; Engel- Cox et al., 2006; Papayannis et al., 2007a,b; Pitari et al., 2013). This study presents the integrated use of satellite remote sensing, sunphotometers and LIDAR for monitoring air pollution in Cyprus.

3. Resources

3.1. CIMEL Sunphotometer

At the main study site in Limassol, the sunphotometer observations were performed by a CIMEL sun-sky radiometer, which is part of the AERONET Global Network (http:// aero- net.gsfc.nasa.gov). The CIMEL sunphotometer allows for measurements of direct solar irradiance and sky radiance at 8 wavelengths; 340, 380, 440, 500, 670, 870, 1020 and 1640 nm. The technical specifications of the instrument are given in detail by Holben et al. (1998).

Figure 1. CUT-TEPAK AERONET station

The instrument is located on the roof of the building of the Department of Civil Engineering and Geomatics of Cyprus University of Technology (CUT) (34.675°N, 33.043°E elevation: 10 m). The CUT_TEPAK AERONET station is located in the center of Limassol, 500m away from the sea. The sunphotometric station has been in operation since April 2010. Figure 1 features the CUT-TEPAK AERONET Cimel sun-photometer.

Figure 2. MICROTOPS II handheld sunphotometer

3.2. MICROTOPS Sunphotometer

At the study sites in Nicosia, Larnaka and Paphos where CIMEL's data were not available a handheld MICROTOPS II sunphotometer was used in order to retrieve AOT measurements.

The sun-photometer is equipped with five accurately aligned optical collimators and internal baffles to eliminate internal reflections. Microtops II provides AOT and water vapor retrievals at five channels, which are determined using the Bouguer-Lambert-Beer law. In order to achieve measurements with great accuracy, the sunphotometer was mounted on a tripod at the same location each time. To avoid cloud contamination, measurements were taken during cloud-free daylight hours. Figure 2 shows the MICROTOPS II handheld sunphotometer used.

3.3. CUT LIDAR System

For the vertical distribution of aerosols, the LIDAR system located at CUT, in Limassol, Cyprus (34.675°N, 33.043°E, 10 m above sea level) was used. The LIDAR records daily measurements between 08:00 UTC and 09:00 UTC (consistent with MODIS overpass) and provides continuous measurements for the retrieval of the aerosol optical properties over Limassol, Cyprus inside the Planetary Boundary Layer (PBL) and the lower free troposphere, thus providing informa- tion for the load, the size and the sphericity of the aerosols.

The LIDAR transmits laser pulses at 532 and 1064 nm simultaneously and collinear with a repetition rate of 20 Hz. This system is based on a small, rugged, flashlamp-pumped Nd-YAG laser with pulse energies around 25 and 56 mJ at 1064 and 532 nm, respectively. An achromatic beam expander reduces the divergence to less than 0.15 mrad. Elastically backscatter signals at two wavelengths (532nm, 1064nm) are collected with a Newtonian telescope with primary mirror diameter of 200 mm and an overall focal length of 1000 mm. The field of view (FOV) of the telescope is 2 mrad. The mirror and cover plate coatings are optimized for the wavelength range from 532 nm to 1064 nm. A plain cover plate protects the mirrors. Behind the field stop two plano-convex with a focal length of 80 mm output parallel rays. The LIDAR covers the whole range starting at the full overlap of the LIDAR (~300 m) up to tropopause level. Three channels are detected, one for the wavelength 1064 nm and two for 532 nm. The two polari- zation components at 532nm are separated in the receiver by means of polarizing beamsplitter cubes (PBC). A special optomechanical design allows the manual ±45°-rotation of the whole depolarization detector module with respect to the laser polarization for evaluating the depolarization calibration constant of the system. The CUT depolarization LIDAR operates at 532nm and it is possible to rotate the detection box including the polarization beam-splitter cube in order to calibrate the instrument (Freudenthaler et al., 2009). Firstly, the backscattered LIDAR signals (P and S) were recorded using the normal orientation of the LIDAR detection box. For the next two steps, the LIDAR detection box is rotated by ±45º, and the P and S signals are recorded. The operation principal of this method is based on the fact that same amount of energy is sent to P and S channels, at "opposite" directions (Freudenthaler et al., 2009). Photomultiplier tubes (PMTs) are used as detectors at all wavelengths except for the signals at 1064 nm (avalanche photodiode, APD). A transient recorder that combines a powerful A/D converter (12 bit at 20 MHz) with a 250 MHz fast photon counting system (Licel, Berlin) is used for the detection of 532 nm radiation, while only analog detection is used at 1064nm. The raw signal spatial resolution is 7.5 meters. The CUT LIDAR system is featured in Figure 3.

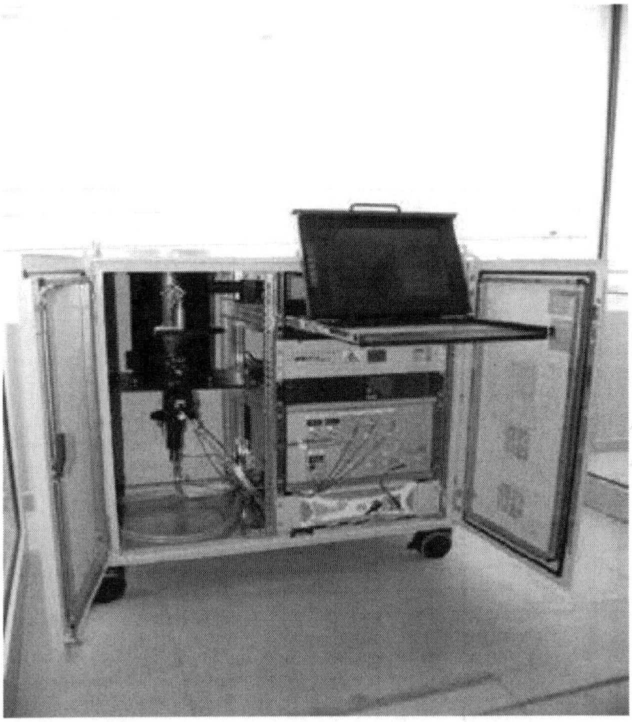

Figure 3. CUT's Depolarization Lidar System

3.4. Surface monitoring

3.4.1. PM_{10} concentration monitoring

Two approached were used to monitor near-surface levels of particulate matter. DustTrack monitors were used at all sites to provide continuous monitoring of PM10. Harvard Impactors were used to collect 24 hour samples of PM10 and PM2.5 which could be analyzed for mass, elemental composition, and other physical-chemical properties of the aerosol. The surface monitoringofparticulatematter(PM)concentrations,TheDustTrack(TSI,Model8533) monitors were located in each of the four sampling sites and were selected to provide weekly monitor- ing of PM_{10} concentrations during morning hours from 08:00 to 13:00 UTC. It records the PM temporal variability with satisfactory time resolution. DustTrak's nominal flow rate of 1.7 l/min is obtained by an internal pump integral to the sampler. The monitor is factory calibrated for the respirable fraction of standard ISO12103-1, A1 test dust (Arizona Test Dust), which is represen- tative of a wide variety of aerosols. It measures concentrations in the range of 0.001– 100 mg/m^3, with a resolution of 0.1% of the reading or 0.001 mg/m^3. Before each measurement, the instru- ment is zeroed and its flow rate is checked. PM_{10} concentrations have been recorded continuous- ly since March 2011. The instrument is located, on the roof of the Cyprus International Institute (CII) in Limassol, at 10 m above ground level in order to avoid the measurements being affected by localized pollution such as passing cars. PM_{10} concentrations were also recorded by Dust- Trak (TSI, Model 8520) at Nicosia, Larnaca and Paphos. One TSI DustTrack has been operated by Frederick University since July 2011 and is located at the top of the Frederick University library building in Nicosia, at 10 m above ground level. Thesecond DustTrack has been operated by CUT's scientific team during 15-day campaigns at Larnaca and Paphos. All sampling points were selected to ensure exposure to wind and to be free of other obstacles. Figure 4 features the TSI Dust Trak. Harvard Impactors were operated each third day at the primary sampling site in Limassol and every sixth day at the other sampling sites.

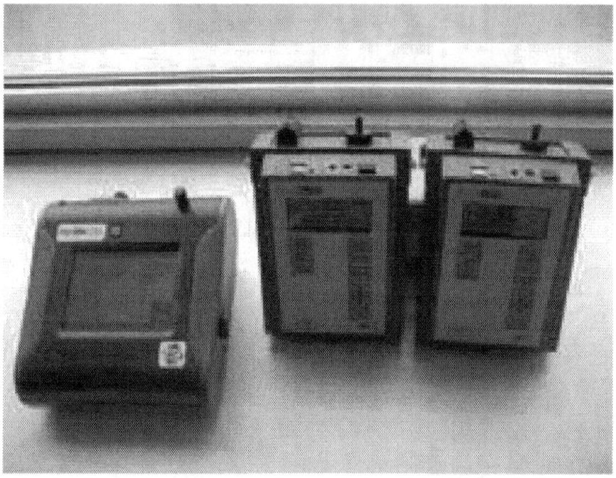

Figure 4. TSI DUST-Track

3.4.2. PM_{10} sampling and elemental composition determinations

Under the AIRSPACE project, the Harvard School of Public Health (HSPH) and Cyprus International Institute for Environmental and Public Health (CII) were responsible

Figure 5. Harvard Samplers

for providing comprehensive and reliable data on the air pollution throughout Cyprus based on ground level measurements.

Air pollution near ground level measurement sites were established in the four cities of Cyprus: Nicosia, Larnaca, Limassol and Paphos. These sites were located at positions thought to be representative of air pollution in each city. In Nicosia, the site is located on the roof of the Frederick University library building, on the same site where the DustTrak and sunphotometer were operated. The Larnaca site is located in the center of the city, on the roof of the tax agency building. The Limassol site is located on the roof of the CII building in the center of the city and Paphos site is on the roof of the economics department of Paphos Municipality. In Figure 5 the setup of the Harvard samplers is presented.

3.4.3. Satellite observations

The Moderate Resolution Imaging Spectro-Radiometer (MODIS) observations from the TERRA and AQUA satellites both measuring spectral radiance in 36 channels (412–14200 nm), in with resolutions between 250 m and 1 km (at nadir) were used to provide a climatology for Cyprus. In polar orbit, approximately 700 km above the Earth, MODIS views a swath of approximately 2300 km resulting in near daily global coverage of Earth's land/ocean/atmos- phere system. The swath is broken into 5-min "granules", each approximately 2,030 km long.

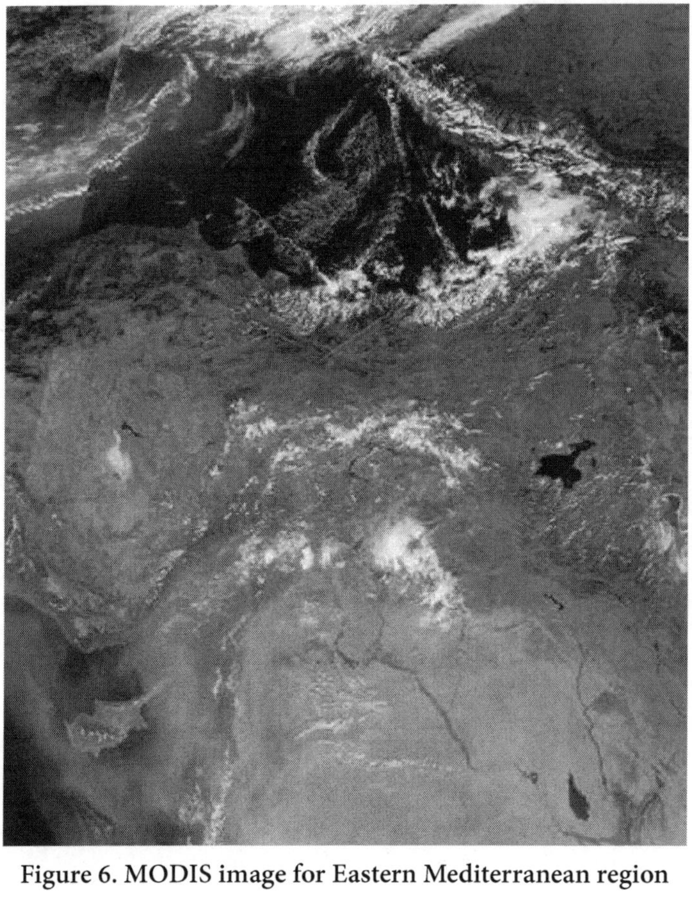

Figure 6. MODIS image for Eastern Mediterranean region

Aerosol products are reported at 10 km resolution (at nadir). Details of file specification of MODIS L2 aerosol products can be found at the website http://modis.gsfc.nasa.gov/.

4. Methodology, study area and data

4.1. Method

The overall methodology is described below (see Fig. 7):

1. **Satellite data products from the MODIS sensor:** Aerosol optical thickness (AOT) and aerosol size/type data were collected for the years 2002-2012 over Cyprus.

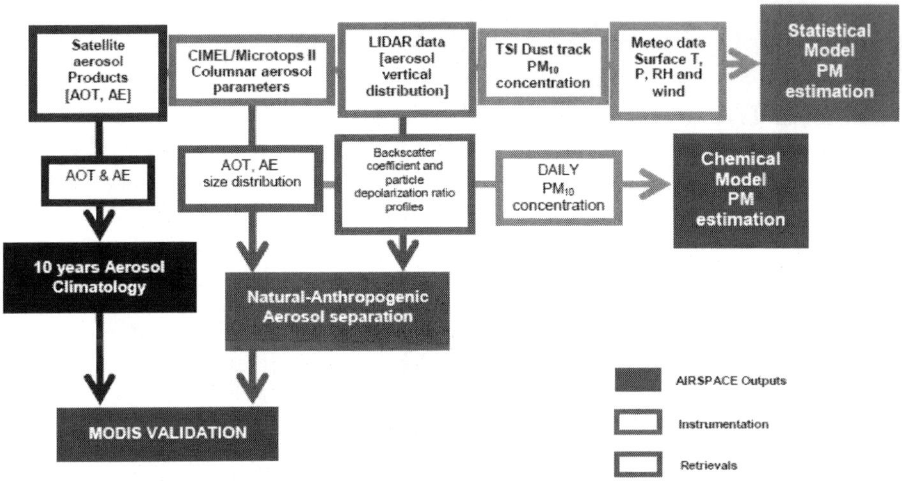

Figure 7. Overall Methodology of the Airspace project

2. **Vertical profile of the aerosol backscatter:** A light detection and ranging (LIDAR) system was established in Limassol in April 2010, consisting of a laser capable of measuring aerosol backscatter and aerosol depolarization ratio in the atmosphere as a function of height. This allows the AOT, and hence the scaling to aerosol concentration, to be quantified below the boundary layer since this fraction best represents the PM measure- ments in a well-mixed boundary layer.

3. **Integrated aerosol optical thickness for the entire atmospheric column:** A sunphotom- eter station was installed in the centre of Limassol (at the CUT premises), where pollution from both industrial and urban sources exist. This further assists in the calibration and verification of satellite derived AOT data. Moreover, two hand-held sunphotometers were used to measure urban, industrial and dust pollution.

4. **Measurements of particulate matter (PM) concentration levels:** Ground level PM was monitored using two methods. Continuous measurements of PM_{10} were taken using portable monitors (DustTrack model 8533). These continuous measurements were supplemented with measurements of PM_{10} and $PM_{2.5}$ taken using Harvard Impactors. The material collected by the Harvard Impactor was analyzed for chemical composition.

5. **Meteorological data from the entire area of Cyprus:** Relative humidity measurements combined with the AOT fraction below the boundary layers, derived by the LIDAR, were incorporated into the statistical PM-AOT models, for improving the PM concentration estimation. Classification of the synoptic situations in Cyprus was also taken into account.
6. **Simulation results from dispersion/air pollution model:** A modeling system that incorporates a fully interactive coupling between the chemistry-aerosol and meteorology (radiation and cloud-physics) portions of the model was created, allowing real-time simultaneous prediction of air quality (in terms of PM aerosol mass) and weather for 72 hours. The model forecasts have been statistically evaluated against surface observations.

4.1.1. Study areas in Cyprus and general characteristics

An overview of the available instrumentations at the selected sites is given in Figure 8.

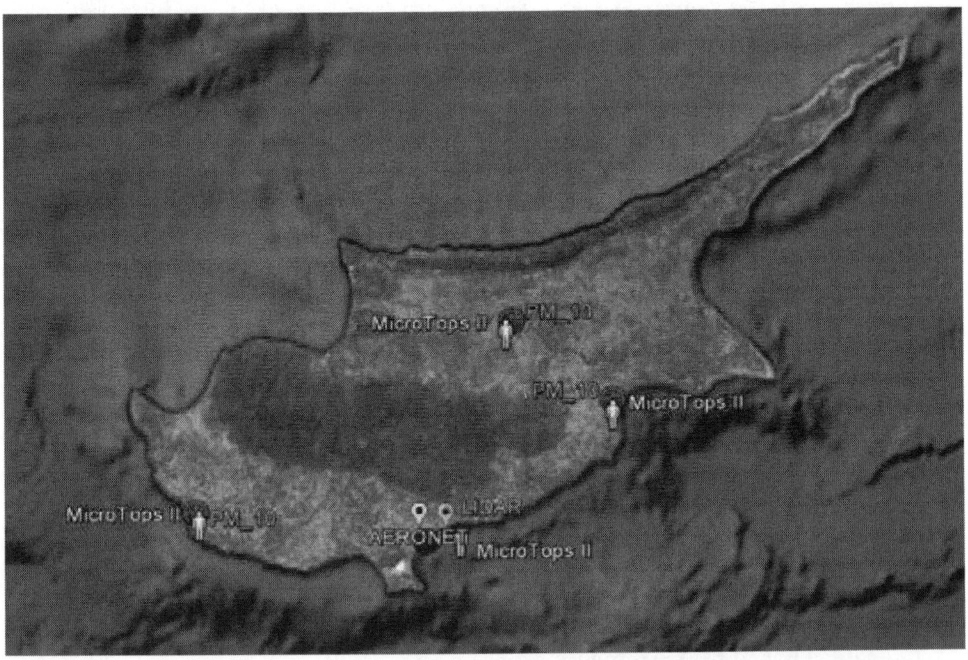

Figure 8. Overview of the available instrumentations at the selected sites within AIRSPACE project. Limassol was the main site (LIDAR, AERONET, PM), Nicosia validation site (MicrotopsII, PM); 15-day campaigns were conducted at Larna- ca and Paphos.

Meteorological conditions: Cyprus is characterized by a subtropical - Mediterranean climate with very mild winters (mainly in the coastal areas) and hot summers. Snowfall occurs mainly in the Troodos Mountains in the centre of the island. Rain occurs mostly during the winter period, with summer being generally dry. Temperature and rainfall are both correlated with altitude and, to a lesser extent, distance from the coast. The prevailing weather conditions on the island are hot, dry summers (from mid-May to mid-

September) and rainy, rather change- able winters (from November to mid-March). These are separated by short autumn and spring seasons.

During the summer period (a season of high temperatures with almost cloudless skies), the island is often under the influence of a shallow trough of low pressure extending from the great continental depression centred over Western Asia. During winter, Cyprus is mainly affected by frequent small depressions traversing the Mediterranean Sea from west to east between the continental anticyclone of Eurasia and the generally low pressure belt of North Africa. These depressions result in disturbed weather usually lasting no more than a few days and producing most of the annual precipitation (the average rainfall from December to February is typically about 60% of the average annual total precipitation). Relative humidity averages between 60% and 80% during the winter period and between 40% and 60% during the summer period. Fog is infrequent and visibility is generally very good. Sunshine is abundant all year round, particularly from April to September when the average duration of bright sunshine exceeds 11 hours per day. Winds are generally light to moderate with high variability when it comes to direction. Gales are infrequent over Cyprus and are mainly confined to exposed coastal areas as well as areas at high elevation.

Aerosol sources: Two main types of air pollutant sources can be identified: anthropogenic and natural. Notable natural sources include dust from inland wind erosion, transboundary sources and sea salt. Cyprus' arid climate results in large portions of surface area having very low index of vegetative cover. This, combined with very low levels of moisture for a substantial part of the year, results in the overall vulnerability to wind erosion. Furthermore, Cyprus presents a high ratio of shoreline when compared to surface area, with maximum distances inland from the shore being in the order of 30-40 km and three of the four urban centres located on the coast. Therefore, sea salt can have a significant effect on the concentrations of particu- lates in the majority of the island's area. Finally, the transportation of dust from the surround- ing eastern Mediterranean and African areas (most notably from northern Africa) significantly affects air quality (Nisantzi et al, 2012).

Local anthropogenic sources also contribute to PM concentrations on the island. The main anthropogenic PM sources include traffic (both highways and inner city traffic), industrial zones, urban agglomerations, agriculture, mines and quarries and localized emissions from a series of activities such as power stations and cement factories.

4.2. The dataset

4.2.1. Ground based measurements

For the purposes of the project, Limassol was selected as the main ground based site for the development and the application of the AIRSPACE methodology. The main instrumentation used for the aerosol observation in a daily basis was a backscatter-depolarization LIDAR system for the study of the vertical aerosol distribution as well as the sunphotometer for the columnar aerosol information, both located at the premises of CUT, in Limassol (see Figure 9) (34.675°N, 33.043°E, 10m above sea level), since 2010. The LIDAR records daily measure- ments between 08:00 UTC and 09:00 UTC (consistent with the MODIS overpass) and performs continuous measurements for the retrieval of the aerosol optical properties such as depolari- zation ratio and backscatter coefficient over Limassol, inside the Planetary Boundary Layer (PBL) and the lower free troposphere.

Additionally, the AERONET sun-photometer provides daily aerosol information including AOT and aerosol size distribution.

Figure 9. Satellite image of Limassol

For the purposes of the AIRSPACE project, Nicosia was selected as a validation site (in addition to the Limassol main site), for ground based measurements of PM_{10} and AOT. Two locations in Nicosia were used as test sites: Strovolos municipality building (N35.144°, 33.343° E) during the period September 2011 to December 2011 and Pallouriotissa Frederick University Research Centre building (N35.181°, 33.379° E) during the period February 2012 to June 2012 and the period October 2012 to January 2012. The Strovolos area is mainly commercial with heavy traffic at peak hours while the Pallouriotissa site is residential.

For both sites, a TSI Dust Trak model 8520 was used for measuring the mass concentration of particulate matter of diameter less than 10 micrometers (PM_{10}). The Dust Trak is a light scattering laser photometer which determines PM_{10} concentrations by measuring the amount of scattering light, which is proportional to the volume concentration of aerosols, in order to determine the mass concentration of aerosols (Nisantzi et al., 2012). The Dust Track features an integrated pump, internal memory and data-logger for automatic storage of measured values at programmable intervals. The device was programmed to begin PM_{10} recordings every morning at 08:00 UTC for a 5-hour period to coincide with the satellite MODIS TERRA and AQUA overpass except at weekends.

Adjacent to the Dust Trak, a Microtops II model 540 sunphotometer was set up to measure the AOT. This is a 5-channel hand-held sunphotometer which measures and stores data at 5 different wavelengths. In addition to the Dust Track and the sunphotometer which were set up originally at the Strovolos site and then moved to the Pallouriotissa site, the Harvard Impactors were operated at the Pallouriotissa site only (next to the other two devices) for chemical analysis of PM_{10}, $PM_{2.5}$, EC-OC and nitrate concentrations.

The in-situ data were collected in conjunction with satellite data (MODIS) to validate a novel statistical model developed within AIRSPACE using AOT retrievals to estimate air particulate pollution.

For Larnaka, two sets of measurements took place: one using the Dust Track along with the Sun photometer for a period of three weeks in August of 2011 (8th-26th) on a site at the centre of Larnaka city (34.916° N, 33.630° E), for the first set of measurements: PM_{10} recordings every morning at 08:00 UTC for a 5-hour period and subsequent measurements using the MICRO- TOPS sun photometer at 08:00 UTC and at 11:00 UTC to coincide with the MODIS TERRA and AQUA overpasses. A second set of measurements was provided by the Harvard Impactor situated on top of the tax agency building (34.919° N, 33.631° E) in Larnaka. This station provided measurements of PM_{10}, $PM_{2.5}$, EC-OC (elemental & organic carbon) and nitrate concentrations.

For air pollution ground level measurements, the Harvard Impactor stations were established by HSPH and CII: Limassol, Nicosia, Larnaca and Paphos. The sampling commenced on 12 January 2012 and ended on 12 January 2013. Samples were collected every six days, on 24-hr basis from 08:00 to 08:00 next day (UTC), at all sites except Limassol, where the sample collection was done every three days. Samples were collected for $PM_{2.5}$, PM_{10}, EC-OC and nitrates using the Harvard Impactors. For quality assurance and control, collocated and blank samples were collected for each sample at the Limassol site, according to a predetermined schedule. Standard Operating Procedure (SOP) was followed for each measurement at each site. Filters were collected and sent to HSPH for chemical analysis. The parameters measured included fine particles ($PM_{2.5}$): mass, reflectance, nitrate, trace elements and EC-OC; and inhalable particle (PM_{10}): mass, reflectance and trace elements. Chemical analysis included Thermal Optical Transmitance (TOT) to measure EC-OC particle concentration, gravimetric mass determination and X-Ray fluorescence to determine trace elemental composition of $PM_{2.5}$ and PM_{10}. Samples up to 19 June 2012 have been completely analyzed. The remaining samples have undergone chemical process for analysis.

5. Results

As described previously, Limassol was the main site for the development of the AIRSPACE methodology for the estimation of the PM levels. The ground based data were used to validate the satellite data. Complementary to the Limassol site, Nicosia's and Larnaca's site observa- tions were used to validate the performance of the models. In this section, the major results from the AIRSPACE project are analysed in some detail.

5.1. Dataset validation

In the AIRSPACE project, both ground based and satellite observations were used to provide aerosol related information for South Eastern Mediterranean region. The first goal of the AIRSPACE project was the validation of the satellite observations in Cyprus, an area affected by aerosol from a variety of sources and surrounded by sea. The ground based observations performed over Limassol and Nicosia were used as the main sites for the validation of the satellite observations.

To incorporate both the spatial and temporal variability of aerosol distribution, the MODIS retrievals at 10 km x 10 km resolution and the AERONET direct Sun measurements at 15-minute intervals (Holben et al., 1998) need to be co-located in space and time.

The AERONET data provide the ground truth for the MODIS validation. The global CUT- TEPAK ground-based AERONET sunphotometer measures aerosol optical thickness in eight channels (340 to 1640 nm). The instrument takes measurements every 15 minutes. From the observations taken within ±30 minutes of MODIS overpass time (Ichoku et al., 2002), mean values of the optical parameters were calculated. Therefore, the maximum number of AERONET observations within the hour of an overpass is 5. Fewer observations within the hour indicate data have been removed by the AERONET Run-Time Cloud Checking procedure.

The study required at least 2 out of possible 5 AERONET measurements to be within ±30 min of MODIS overpasses and at least 5 out of possible 25 MODIS retrievals to be within a 25 km radius centred over the AERONET site. The mean values of the collocated spatial and temporal ensemble were then used in a linear regression analysis and in calculating RMS errors. The AERONET level 1.5 data were cloud screened. Though the level 2.0 data provide final calibration, they are not available for the entire time period of the project. Therefore, the level 1.5 data (instead of level 2.0) were used in the operational MODIS aerosol validation scheme.

A total of 352 points of AERONET site representing the correlated criteria for the MODIS- and AERONET derived AOT were collected in the period from April 2010 to December 2012.

Figure 10 features the correlation of the MODIS AQUA and TERRA sensors and CUT_ TEPAK AERONET measurements. The slope of linear regression in the correlation plot between MODIS and AERONET provides an overview of possible differences. The correlation coeffi- cient value of the order of 0.62 for both TERRA and AQUA satellites is due to the coast line of the Limassol site. Limassol's CUT-TEPAK AERONET site is a coastal area, thus the surface inhomogeneity or sub-pixel water contamination has a larger effect than anticipated in continental coastal regions (Nisantzi et al., 2012). The systematic biases overestimations in MODIS retrievals are mainly due to aerosol model assumptions (deviation of 0–20%) andin- strument calibration (2–5%).

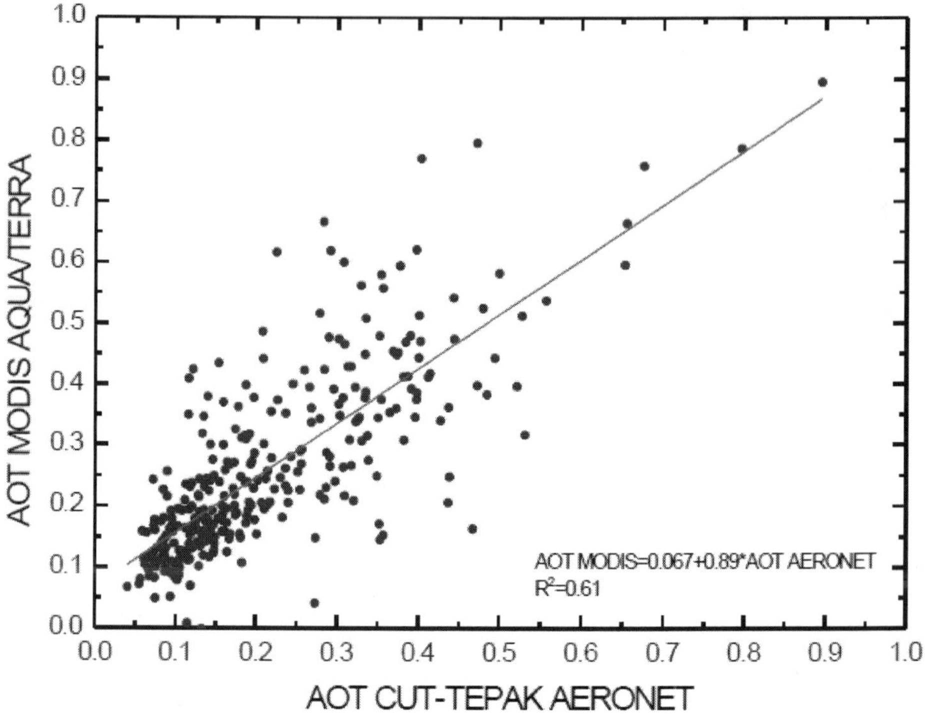

Figure 10. Comparisons of MODIS and AERONET derived at 0.50 nm wavelength, encompassing 352 points from CUT- TEPAK AERONET coastal site. The solid line represents the slopes of linear regression both for AQUA and TERRA MODIS sensors

Using the MICROTOPS II AOT, the procedure was duplicated for the validation of the satellite observations in Nicosia. The number of collocated and synchronized ground based and satellite measurements were statistically low in order to provide correlation factor which can represent a reliable validation study.

5.2. Satellite climatology

In the present work, the Level 2, 10x10km, MOD04 aerosol products (Collection 051) were retrieved for the years 2001 to 2011 from NASA's Level 1 and Atmosphere Archive and Distribution System (LAADS). The AOT fields were extracted from the 'Opti- cal_ Depth_Land_And_Ocean' parameter which provides the AOT at 550nm derived via the dark-target algorithms and with best quality data (Remer et al., 2005). According to Remer et al. (2009), the AOT fields for this product have been respectively validated to within the error bounds of (0.04+0.05AOT) and ±(0.05+0.15AOT) at 550nm.

Based on the above AOT data, subsets for the area of Cyprus were extracted and mean monthly climatology maps were constructed for the period 2001-2011. For the area considered, the number of days with valid TERRA AOT measurements ranged approximately from 1000 to 2300 (which amount to 25%-57% time coverage), as shown in Figure 11. The highest number of valid measurements was observed over the central area of Cyprus (in the vicinity of Troodos Mountain), whereas near the coastline, this number decreased.

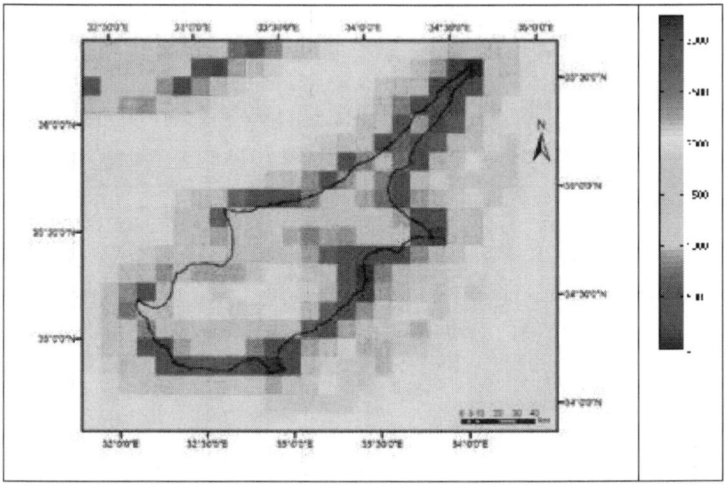

Figure 11. Number of valid TERA AOT observations for the period 2001-2011

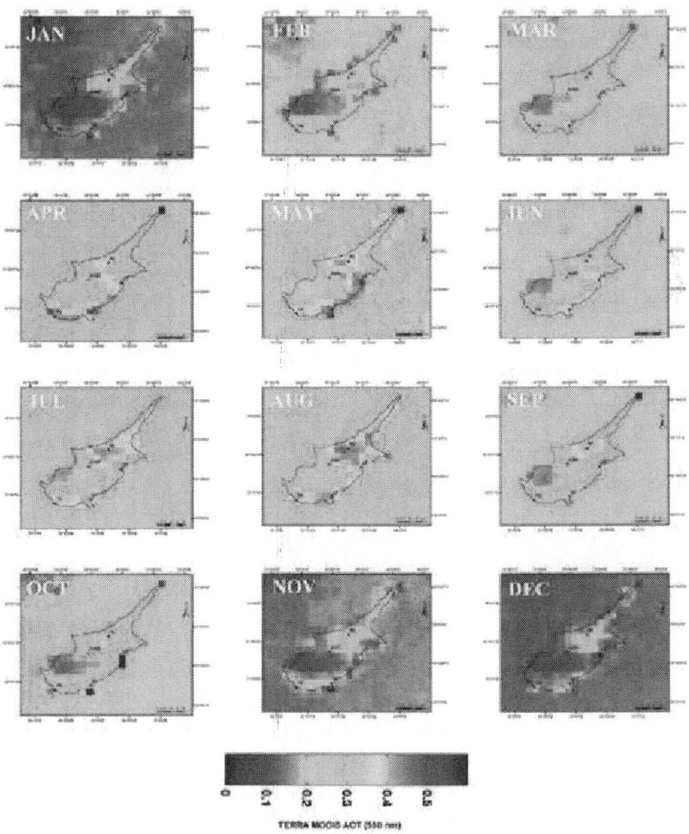

Figure 12. Average monthly AOT. (LE, LA, LM, and AM mark the sites of Nicosia, Larnaca, Limassol and Ag. Marina)

The maps for each month are presented in Figure 12. The seasonal cycle of the aerosol load is well depicted. Minima are observed during winter months and maxima during spring and summer when intense phenomena associated with dust transport from Sahara desert are more frequent. The respective monthly average values for the three urban sites of Nicosia, Larnaca, and Limassol (marked as LE, LA and LM, respectively, on the maps) and the background site of Agia Marina, (marked as AM) have been calculated. In general, the background site is characterised by lower aerosol loads (ranging from 0.1 to 0.28) than those observed at the urban sites. Limassol (the main port city) presents the highest values for the period January-May and Nicosia (the capital city) from June to December. For this latter period, Larnaca presents intermediate values. The two distinct maxima associated with dust transport phenomena are observed at all sites in May and August. The value for the first peak in May is approximately the same for all urban sites (~0.40) but for August, the levels for Nicosia are higher (~0.45) compared to the other two urban sites (~0.35 for Larnaca and Limassol).

5.3. PM surface analysis

One element of the AIRSPACE program in Cyprus was the measurement of ground level PM concentrations by Harvard Impactors.

Statistics for the Limassol site show that for the first six months of observations the mean value for PM_{10} is 32.1 μg/m^3, for $PM_{2.5}$ 13.4 μg/m^3 and for total carbon 2.3 μg/m^3 with standard deviations of 20.9, 4.6 and 1.1 μg/m^3, respectively.

PM_{10} and $PM_{2.5}$ were analysed, for trace elements such as sulfur, magnesium, aluminum, sodium, silicon, chlorine, potassium and calcium. Statistics for some of those trace elements for $PM_{2.5}$ are shown below, in Table 1.

Table 1. Trace elements statistics for the first 6 months sampling, for $PM_{2.5}$

	Mean (μg/m3)	SD (μg/m3)	Median (μg/m3)
Sodium	0.27	0.15	0.23
Magnesium	0.06	0.08	0.05
Aluminum	0.16	0.29	0.07
Silicon	0.29	0.55	0.13
Sulfur	1.29	0.73	0.99
Chlorine	0.07	0.17	0.02
Potassium	0.11	0.06	0.10
Calcium	0.20	0.37	0.12

Figure 13 and 14 indicate the 6 month time series of the $PM_{2.5}$ and PM_{10} concentrations, as well as the elemental, organic and total carbon levels from the Limassol filters.

Analysis of these initial samples revealed evidence of a dust storm event recorded on 12 March 2012, with PM_{10} and $PM_{2.5}$ concentrations reaching up to 156.6 μg/m³ and 29.4 μg/m³, respec- tively. These values are several times higher than the typical values shown during the sampling period and well above the 24-hour limit value set by EEA, especially for PM_{10}.

PM10 and PM2.5 concentrations show a small increase from the start of the sampling (Janu- ary 2012) until June 2012, indicating a temporal relationship.

5.4. Statistical model

Based on the data collected a statistical model was established for estimation of PM concen- trations from AOT measurements. Using a general linear regression model, the AOT retrieved by MODIS was used to predict ground-level PM_{10} concentrations in Limassol, Cyprus.

The proposed model by Liu et al. (2007) is given in equation 1:

$$n(PM_{10}) = \beta_0 + \beta_1(\log AOT) + \beta_2(\log AE) + \beta_3(WVdep) + \beta_4(\ln(T)) + \beta_5(\ln(RH)) + \beta_6(\ln(WS)) + \beta_7(Wd) + \beta_8(P) + \beta_9(PBL) \quad (1)$$

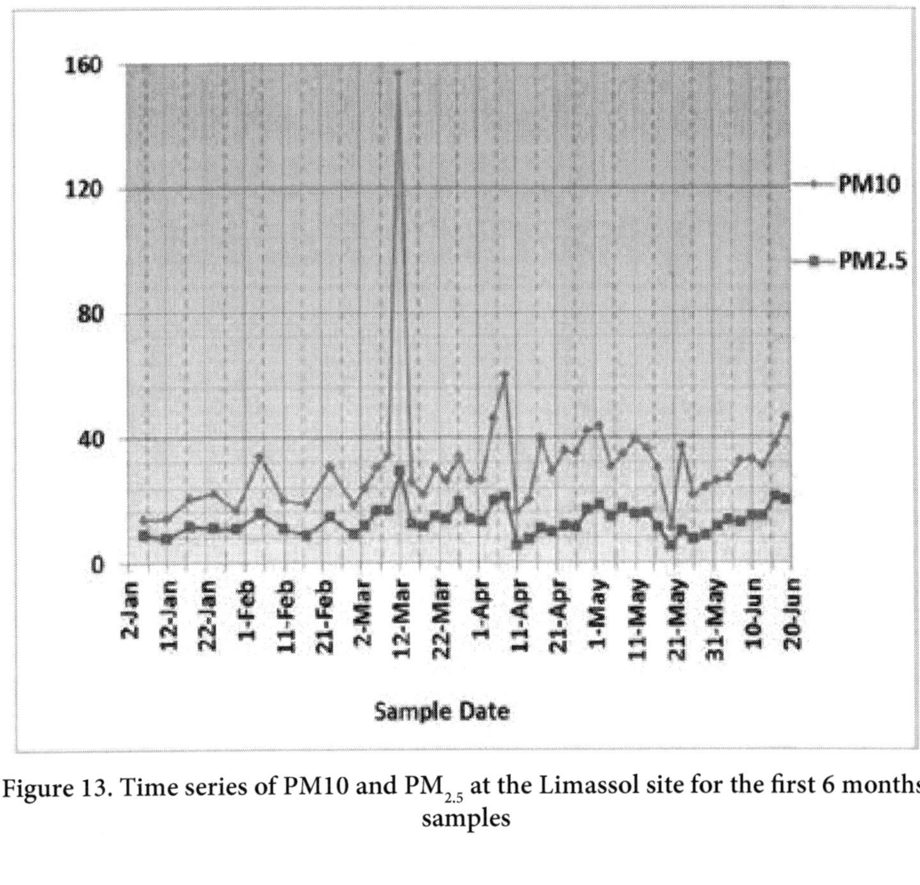

Figure 13. Time series of PM10 and $PM_{2.5}$ at the Limassol site for the first 6 months' samples

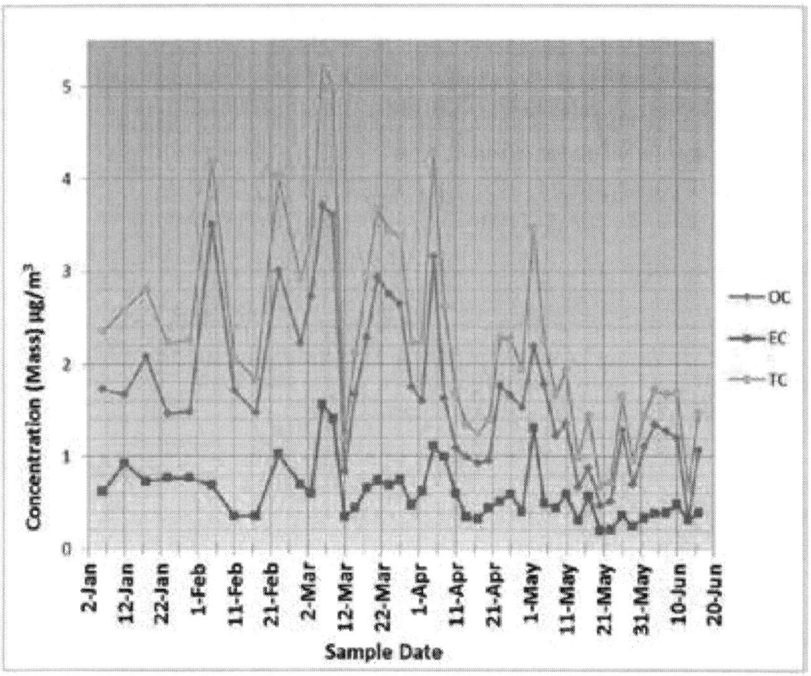

Figure 14. Time series of organic (OC), elemental (EC), and total carbon (TC) at the Limassol site for the first 6 months' samples.

Where βi are the regression coefficients, AOT is the Aerosol Optical Thickness, AE is the Ångström Exponent, WV is the Water Vapour, T is the surface temperature, WS is the wind speed, Wd the wind direction, P is the pressure at surface level and PBL is the Planetary boundary layer height.

The available data set in AIRSPACE project are given in Table 3:

Table 2. AIRSPACE dataset used for the statistical model

Parameters	Instrument
Aerosol Optical Depth	CIMEL
Angstrom Exponent	CIMEL
Total Column Water Vapour	CIMEL
PM 10	Dust Track TSI
PBL height	LIDAR
Meteorological Data	METAR-LCRA (Akrotiri Air Base, Cyprus)

Based on the proposed methodology, the performance of the multi-regression model was examined by introducing one predictor (Xi) at a time, together with the initial predictor, the AOT at 500nm (Xi i=0). For each predictor Xi, four transformations (j) were considered

Table 3.

# transformation	Type of parameter involved
1	Ln(Xi)
2Xi	
3	Departures from mean value of Xi
4	Ratio of mean value of Xi

From the above options (j=1 to 4), the one with the highest correlation coefficient (CCij) between predicted and measured PM_{10} was selected. In each iteration step k, the maximum values of the CCij = CCik were compared, in order to select the predictor Xik with the highest positive impact. Due to the limited dataset, no evident seasonal dependence was noted (Cook and Sanford, 1982).

The results are presented below. In Figure 15 the correlation coefficient between the predicted and measured PM_{10} is presented for 8 different models. The maximum performance of the model is reached by using the following predictors (in strength order), with a correlation coefficient on the order of CC=0.85

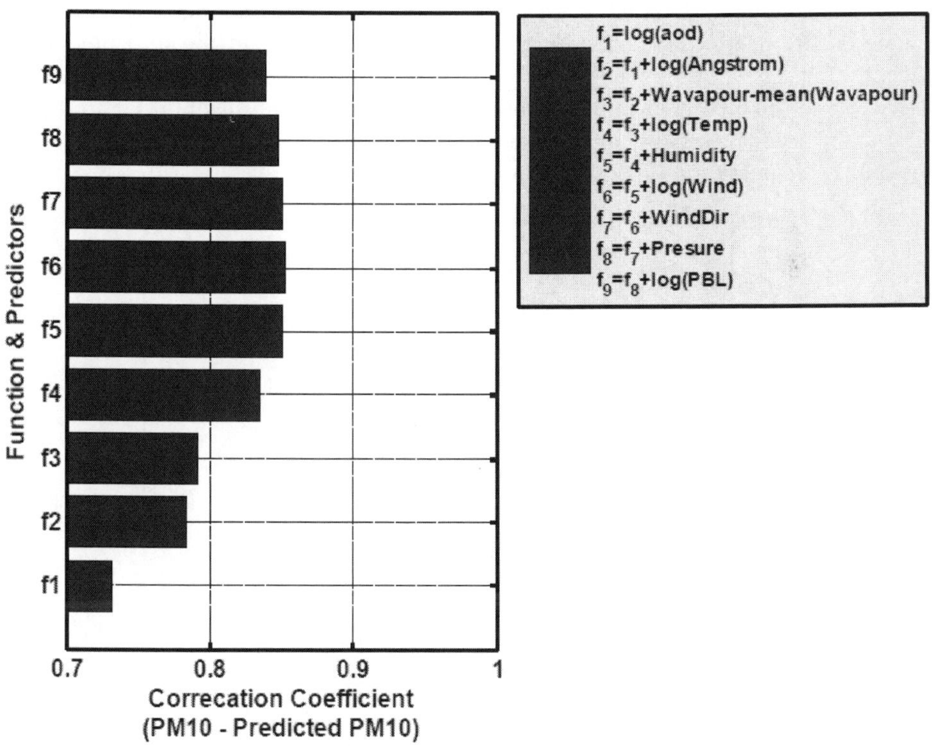

Figure 15. The correlation coefficient between the predicted and measured PM10 is presented for 8 different models

Table 4. Best correlation coefficients and regression coefficients

	CC= 0.853	CC=0.850
Constant Term	-4.117	-4.111
Ln(AOT)	0.952	0.943
Ln(AE)	0.299	0.299
Wavapour-mean(Wavapour)	-0.393	-0.384
Ln(Temp)	0.643	0.6.23
Humidity	0.008	0.008
Ln(Wind)	-0.096	-0.071
WindDir		-0.0004

The results of the above sensitivity analysis indicate the maximum performance of the model of the order of CC=0.85 as shown in equation 2:

$$Ln(PM_{10}) = -4.11 + 0.952(\log AOT) + 0.299(\log AE) - 0.393(WVdep) + 0.643(\ln(T)) + 0.008(RH) - 0.096(\ln(WS)) \quad (2)$$

Finally, using formula 2 as the best model and the coefficients derived and shown in Table 4, the relationship between the model's prediction and the measured PM10 concentrations, is shown in Figure 16. The residuals, i.e, the differences between the measured and the predicted values of the PM concentration are shown in Figure 17. The points in the residual plot in Figure 17 are randomly dispersed around the horizontal axis, thus, a linear regression model is appropriate.

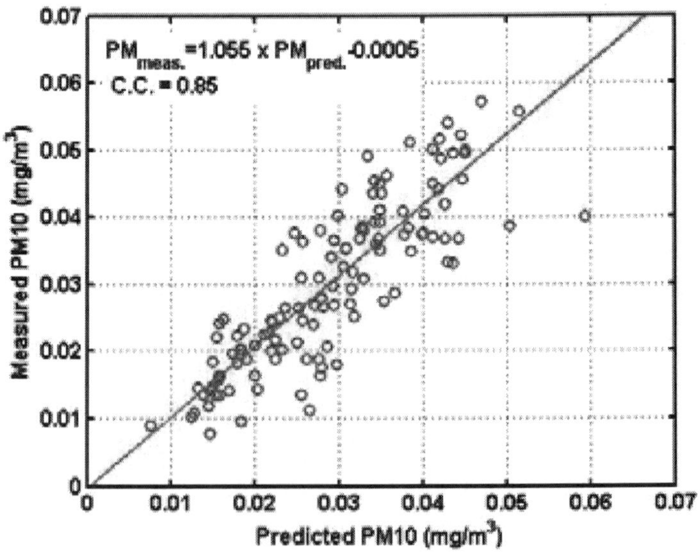

Figure 16. Comparison between predicted and measured PM10 by TSI DUST Track at Limassol (Red line : linear fit)

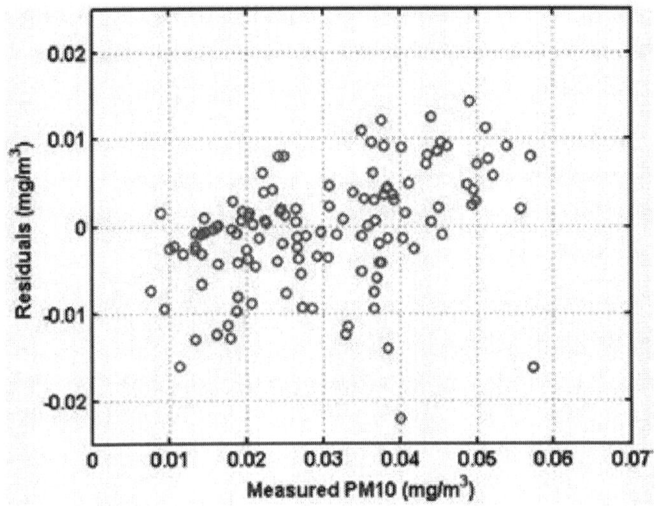

Figure 17. Differences between the measured and the predicted value of the PM concentration (Residual plots)

5.5. Chemical model

Within AIRSPACE project, a high resolution atmospheric Chemistry General Circulation Model (AC-GCM) was used to study the emission, transport and deposition of dust. The Modular Earth Sub-model System (MESSy version 2.41) (Joeckel et al., 2005; 2006; 2010) is an earth system model which is capable of running with multiple representations of processes simultaneously paired to the core atmospheric general circulation model (ECHAM5). The model configuration used in the present study has a spectral resolution of T255L31 (0.5°, 50km) and 31 vertical levels up to 10 hPa. Gleser et al. (2012) emphasized the importance of higher resolution simulations for better dust representation in the model. As this is a global model, no boundary conditions are necessary. All known emission sources are included, while the initial conditions originate from the ERA40 reanalysis data (European Centre for Medium- Range Weather Forecasts - ECMWF) at 0.5-degree resolution. Every 12 hours of operation, the model fields are moved towards the ERA40 data in order to simulate the meteorological conditions, as precised as possible. In order to reduce computational time, the model uses a simplified chemistry module, preserving only the sulfate and NOx interactions which are considered the most important as far as the aerosols are considered. The model output is averaged and stored over 5-hour intervals, which provides an entire diurnal cycle after 5 days. The configuration includes also a simplified sulphate chemistry scheme (Gleser et al., 2012) allowing the production of sulphuric acid and particulate sulphate, which play an important role in transforming dust particles from hydrophobic into hydrophilic, thus affecting their ability to interact with clouds and be removed by precipitation (Astitha et al., 2012). The ammonia (NH_3) reaction with sulphate and corresponding coating with dust (Ginoux et al., 2012) is also considered in this study. Due to the focus on dust episodes, a reduced version of the atmospheric chemistry scheme was applied which did not account for secondary inorganic and organic aerosol species associated with air pollution. The model used ECMWF gridded meteorological data to represent the actual meteorological conditions. To ensure adequate representation of the pollutants and dust in the atmosphere, the model runs for 15 days (spin- off) to create from the meteorology and the emissions the current weather conditions. This strategy ensures that the existing pollutants not represented in the model are removed from the atmosphere, while the sources will produce pollutants that will be dispersed in the atmosphere. After the initial spin-off, the atmospheric conditions represented from the model fields and the pollutant concentrations are considered as close to reality as possible. The model simulation was performed over the period of September to October 2011.

The most significant issue for the operational run of a numerical model prediction of the dust is the complete absence of initial conditions for pollutant and dust concentrations. This enforces the utilization of global models to simulate the atmosphere with extremely accurate emission inventories which are absent or not complete for North Africa and Eastern Mediter- ranean. The latter is an important source of uncertainty for concentrations. Furthermore, the sparse coverage of measurements for the spatial validation of the model in the region does not provide a clear picture for the evaluation assessment of the model.

The use of a global model necessitated the utilization of a large grid due to computational limitations. The global grid introduced an adequate representation of the topography of the models and requires special parameterization of processes that often lead to errors. Another restriction is the simplified chemistry used for the simulation. The computational power necessary for the implementation of a full chemistry scheme is not currently available.

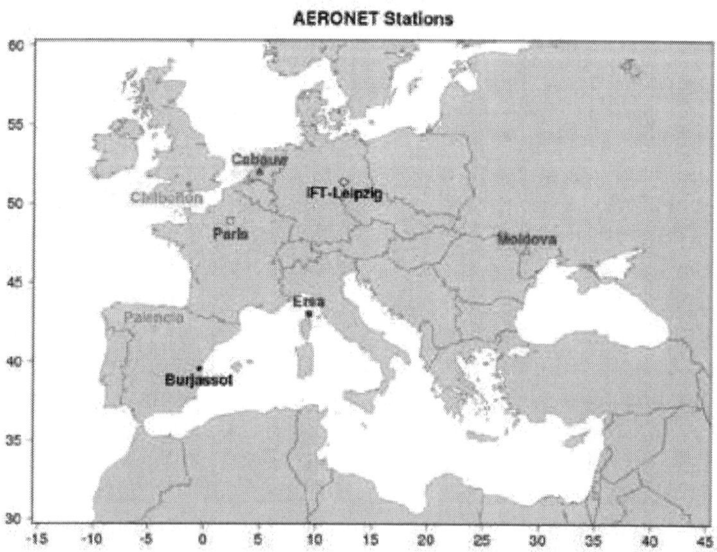

Figure 18. AERONET stations used to evaluate the model results

The model results were evaluated using the AOT fields provided by the NASA AERONET available from http://aeronet.gsfc.nasa.gov. The data comparison represents the AOT for all aerosols simulated in the model as well as those observed in the atmosphere at 550nm wavelength. The observed AOT was averaged over the 5-hour output intervals in line with the averaged AOT over the same period from the model. Figure 18 shows the eight AERONET stations which observational data were available during the simulation period and which were used in this study. These stations are not necessarily located in dust-dominated regions but can be more strongly affected by other aerosol types, including air pollution.

The scatter plot between the modeled and observed AOT is shown in Figure 19. Different colors and symbols are used for each station ID (see legend). As shown, the model is capable of simulating the AOT in general. However, at some stations (Leipzig, Palencia, Paris) the model tends to underestimate the observed AOT. This is explained by the use of the reduced atmospheric chemistry scheme in the model that does not fully account for urban air pollution in addition to the unresolved physics at small scales in the global models. However, the comparison of the output of the model for the AOT with the measured values from the AERONET network indicates that the simulated atmosphere is valid in areas with similar climatological and industrial characteristics to Cyprus, while for areas with heavy industry, there is a significant deviation which can be justified from the reduced chemistry module used for the runs.

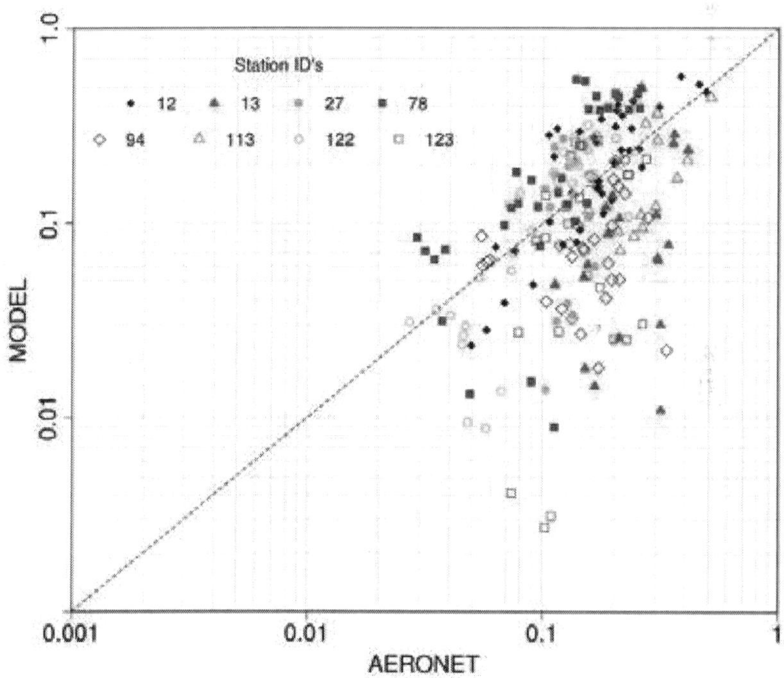

Figure 19. Scatter plot between modeled and observed AOT for different AERONET stations

Furthermore, model AOT estimations have been compared with the available AOT measure- ments from CUT-TEPAK AERONET site. Figure 20 shows the time evolution of the AOT for the Limassol AERONET station together with the model results. As shown in Figure 20, the model is generally, in agreement with observations in both magnitude and timing for Limassol with respect to the average measured values. The comparison between the modeled and observed AOT indicates the ability of the model to simulate the AOT adequately.

6. Conclusions

An integrated methodology for assessing and studying air pollution in several areas of Cyprus was presented through the AIRSPACE project. Satellite derived aerosol optical thickness data along with LIDAR, sun-photometric and in-situ (PM) measurements were analyzed. The proposed integration of several tools and technologies provides to the user an alternative way for assessing and monitoring air pollution.

First, a new multiple linear regression model for estimating PM_{10} using AOT values and some other auxiliary meteorological atmospheric parameters has been developed for the urban area of Limassol in Cyprus. AOT can be retrieved by satellite sensors and is validated on the ground by using measured values with sunphotometers. Such model can be used for future satellite acquisitions. The integrated use of several resources and technologies such as satellite image data, LIDAR measurements, meteorological data and sunphotometric data lead to the development of new approaches in estimating PM concentrations.

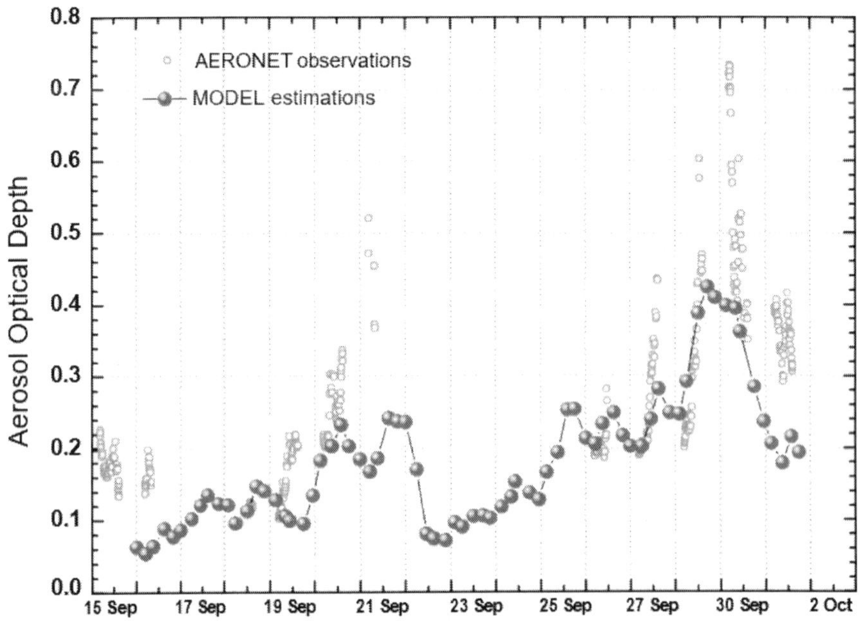

Figure 20. Comparison between the modeled (dark green) and observed (light green) AOT for Limassol AERONET on September 2011

Second, an atmospheric chemical simulation model was run for the period September-October 2011. The model results were evaluated using the AOT provided by the NASA AERONET. AOT estimations have been compared with the available AOT measurements from CUT- TEPAK AERONET site. It has been found that the modeled and observed AOT values were in good agreement, except during the periods of peak PM concentrations.

References

1. Astitha, M, Lelieveld, J, Abdel, M, & Kader, A. Pozzer, and A. de Meij. "New parameterization of dust emissions in the global atmospheric chemistry-climate model EMAC", Atmospheric Chemistry and Physics Discussions, 12(5):13237{13298, May (2012).

2. Bosenberg, J, et al. (2003). EARLINET: A European Aerosol Research LIDAR Network, Rep. 348, MPI-Rep. 337, 191 pp., Max-Planck-Inst. fur Meteorol., Hamburg, Germany.

3. Chudnovsky, A, Kostinski, A, Lyapustin, A, & Koutrakis, P. Spatial scales of pollu- tion from variable resolution satellite imaging", Environmental Pollution, (2013). , 172, 131-138.

4. Cook, R. Dennis, and Sanford Weisberg. Residuals and influence in regression. Vol. 5. New York: Chapman and Hall, (1982).

5. Engel-cox, K, Hoff, R, Rogers, R, Dimmick, F, Rush, A, Szykman, J, Al-saadie, J, Chu, A, & Zell, E. Integrating lidar and satellite optical depth with ambient monitoring for 3-dimensional particulate characterization", Atmospheric Environment, (2006). , 40, 8056-8067.

6. Freudenthaler, V, Esselborn, M, & Wiegner, . , M., Depolarization ratio profiling at several wavelengths in pure Saharan dust during SAMUM 2006. Tellus B, 61: 165-179. doi: 10.1111/j.1600-0889.2008.00396.x, 2009

7. Ginoux, P, Clarisse, L, Clerbaux, C, Coheur, P. -F, Dubovik, O, Hsu, N. C, & Van Damme, M. Mixing of dust and NH3 observed globally over anthropogenic dust sources. Atmospheric Chemistry and Physics, 12(16):7351{7363, Aug. (2012).

8. Gleser, G, Kerkweg, A, & Wernli, H. The mineral dust cycle in EMAC 2.40: sensitivity to the spectral resolution and the dust emission scheme". Atmospheric Chemistry and Physics, 12(3):1611{1627, Feb. (2012).

9. Grosso, N, & Paronis, D. Comparison of contrast reduction based MODIS AOT estimates with AERONET measurements", Atmospheric Research, (2012). , 116, 33-45.

10. Gupta, P, Christopher, S. A, & Wang, J. Satellite remote sensing of particulate matter and air quality assessment over global cities", Atmospheric Environment, 40, 5880–5892, (2006).

11. Hadjimitsis, D. G, Agapiou, A, Themistocleous, K, Achileos, C, Nisantzi, A, Panayiotou, C, & Kleanthous, S. Air pollution monitoring based on remote sensing data and simultaneous ground PM 10 and PM2.5 measurements: the WebAir-2 project, 11th International Conference on Meteorology, Climatology and Atmospheric Physics, Athens, Greece 30/(2012). , 5-1.

12. Hadjimitsis, D. G. Description of a new method for retrieving the aerosol optical thickness from satellite remotely sensed imagery using the maximum contrast value principle and the darkest pixel approach, Transactions in GIS Journal, 12(5), DOI:j. 1467-9671.2008.01121.x, (2008). , 633-644.

13. Hadjimitsis, D. G, Themistocleous, K, Nisantzi, A, & Matsas, A. The study of atmospheric correction of satellite remotely sensed images intended for air pollution using sunphotometers (AERONET) and LIDAR system in Lemesos, Cyprus, Proceedings of SPIE V, (2010). , 7832

14. Hadjimitsis, D. G. Aerosol Optical Thickness (AOT) retrieval over land using satellite image-based algorithm, Air Quality, Atmosphere & Health- An International Jour- nal, 2 (2), DOIs11869-009-0036-0, (2009). , 89-97.

15. Hadjimitsis, D. G, Retalis, A, & Clayton, C. R. I. The assessment of atmospheric pollution using satellite remote sensing technology in large cities in the vicinity of airports, Water, Air & Soil Pollution: Focus, An International Journal of Environmental Pollution, 2 (5-6): 631-640. DOI:A:102130541700, (2002).

16. Holben, B. N, et al. AERONET-A federated instrument network and data archive for aerosol characterization, Remote Sens. Environ., 66, 1-16, (1998).

17. Hostetler, C. A, Liu, Z, & Reagan, J. CALIOP Algorithm Theoretical Basis Document, Calibration and Level 1 Data Products, Document No. PC-SCI-201, NASA, (2006).

18. Ichoku, C, Chu, D. A, Mattoo, S, Kaufman, Y. J, Remer, L. A, Tanré, D, Slutsker, I, Holben, B. N, & Spatio-temporal, A. approach for global validation and analysis of MODIS aerosol products, Geophys. Res. Lett., 29(12), doi:10.1029/2001GL013206, (2002).

19. Joeckel, P, Kerkweg, A, Pozzer, A, Sander, R, Tost, H, Riede, H, Baumgaertner, A, Gromov, S, & Kern, B. Development cycle 2 of the modular earth submodel system (MESSy2)", Geoscience Model Development, 3(2):717{752, Dec. (2010).

20. Joeckel, P, Tost, H, Pozzer, A, Bruehl, C, Buchholz, J, Ganzeveld, L, Hoor, P, Ker- kweg, A, Lawrence, M, Sander, R, Steil, B, Stiller, G, Tanarhte, M, Taraborrelli, D, Van Aardenne, J, & Lelieveld, J. The atmospheric chemistry general circulation mod- el ECHAM5/MESSy1: consistent simulation of ozone from the surface to the mesosphere", Atmospheric Chemistry and Physics, 6(12):5067{5104, Nov. (2006).

21. Joeckel, P, Sander, R, Kerkweg, A, Tost, H, & Lelieveld, J. Technical note: The modu- lar earth submodel system (MESSy)- a new approach towards earth system model- ing", Atmospheric Chemistry and Physics, 5(2):433{444, Feb. (2005).

22. Jones, A. J, & Christopher, S. A. MODIS derived fine mode fraction characteristics of marine, dust, and anthropogenic aerosols over the ocean, constrained by GOCART, MOPITT, and TOMS", Journal of Geophysical Research, 112:D22204; doi: 10.1029/2007JD008974,(2007).

23. Koelemeijer, R. B, Homan, C. D, & Matthijsen, J. Comparison of spatial and temporal variations of aerosol optical thickness and particulate matter over Europe", Atmospheric Environment, 40, 5304–5315, (2006).

24. Liu, Y, Franklin, M, Kahn, R, & Koutrakis, P. Using aerosol optical thickness to predict ground-level PM2.5 concentrations in the St. Louis area: A comparison between MISR and MODIS. Remote sensing of Environment, 107(1), 33-44, (2007).

25. Michaelides, S, Tymvios, F, Paronis, D, & Retalis, A. Artificial neural networks for the diagnosis and prediction of desert dust transport episodes", Studies in Fuzziness and Soft Computing, (2011). , 269, 285-304.

26. Middleton, N, Yiallouros, P, Kleanthous, S, Kolokotroni, O, & Schwartz, J. Dockery W. D.,

27. Demokritou, P, Koutrakis, P, & Time-series, A 10y. e. a. r. analysis of respiratory and cardiovascular morbidity in Nicosia, Cyprus: the effect of short-term changes in air pollution and dust storms". Environmental Health (2008). doi:X-7-39.

28. Neophytou, M. A, Yiallouros, P, Coull, A/ B, Kleanthous, S, Pavlou, P, Pashiardis, S, Dockery, W. D, Koutrakis, P, & Laden, F. Particulate matter concentrations during desert dust outbreaks and daily mortality in Nicosia, Cyprus", Journal of Exposure Science & Environmental Epidemiology, 20 February (2013). doi:jes.2013.10.

29. Nisantzi, A, Hadjimitsis, D. G, Agapiou, A, Themistokleous, K, Michaelides, S, Tym- bios, F, Charalambous, D, Athanasatos, S, Retalis, A, Paronis, D, Perdikou, S, Koutra- kis, P, Evans, J. S, & Achilleos, S. Study of air pollution with the integrated use of MODIS data, LIDAR, sun photometers and ground PM sampler measurements in Cyprus, 11th International Conference on Meteorology, Climatology and Atmospheric Physics, Athens, Greece 30/(2012)., 5-1.

30. Papayannis, A, Mamouri, R. E, Chourdakis, G, Georgoussis, G, Amiridis, A, Paronis, D, Tsaknakis, G, & Avdikos, G. Retrieval of the optical properties of tropospheric aerosols over Athens, Greece combining a wavelength Raman-lidar and the CALIPSO VIS-NIR lidar system: Case-study analysis of a Saharan dust intrusion over the Eastern Mediterranean", Journal of Optoelectronics and Advanced Materials, 9 (11), 3514-3517, (2007a)., 6.

31. Papayannis, A, Zhang, H. Q, Amiridis, V, Ju, H. B, Chourdakis, G, Georgoussis, G, Pérez, C, Chen, H. B, Goloub, P, Mamouri, R. E, Kazadzis, S, Paronis, D, Tsaknakis, G, & Baldasano, J. M. Extraordinary dust event over Beijing, China, during April 2006: Lidar, Sun photometric, satellite observations and model validation", Geophys- ical Research Letters, 34 (7), (2007b).

32. Pitari, G. Di Carlo P., Coppari E., De Luca N., Di Genova G., Iarlori M., Pietropaolo E., Rizi V. and Tuccella P., "Aerosol measurements at L'Aquila EARLINET station in central Italy: Impact of local sources and large scale transport resolved by LIDAR", Journal of Atmospheric and Solar-Terrestrial Physics, (2013)., 92, 116-123.

33. Pringle, K. J, Tost, H, Metzger, S, Steil, B, Giannadaki, D, Nenes, A, Fountoukis, C, Stier, P, Vignati, E, & Lelieveld, J. Description and evaluation of GMXe: a new aero- sol submodel for global simulations (Geoscience Model Development, 3(2):391{412, Sept. (2010)., 1

34. Remer, L. A, et al. An emerging aerosol climatology from the MODIS satellite sensors, J. Geophys. Res., 113, D14S07, (2008).

35. Retalis, A, & Sifakis, N. Urban aerosol mapping over Athens using the differential textural analysis (DTA) algorithm on MERIS-ENVISAT data", ISPRS Journal of Photogrammetry and Remote Sensing, (2010)., 65, 17-25.

36. Retalis, A, Sifakis, N, Grosso, N, Paronis, D, & Sarigiannis, D. Aerosol optical thickness retrieval from AVHRR images over the Athens urban area", Proc. IEEE Interna-

tional Geoscience & Remote Sensing Symposium (IGARSS) 2003, 21-25 July 2003, Toulouse, France, Vol. IV. (2003). , 2182-2184.

37. Retalis, A, Cartalis, C, & Athanassiou, E. Assessment of the distribution of aerosols in the area of Athens with the use of LANDSAT Thematic Mapper data". International Journal of Remote Sensing 20 (5), 939-945, (1999).
38. Van Donkelaar, A, Martin, R. V, Brauer, M, Kahn, R, Levy, R, Verduzco, C, & Villeneuve, P. J. Global Estimates of Ambient Fine Particulate Matter Concentrations from Satellite-Based Aerosol Optical Depth: Development and Application", Environmen- tal Health-Perspectives, 118 (6), 847-855, (2010).
39. Vidot, J, Santer, R, & Ramon, D. Atmospheric particulate matter (PM) estimation from SeaWiFS imagery", Remote Sensing Environment, 111, 1–10, (2007).
40. Wald, L, & Baleynaud, J. M., "Observing air quality over the city of Nantes by means of Landsat thermal in frared data". International Journal of Remote Sensing, 20, 5, 947-959, (1999).
41. WHO(2006). Health risks of particulate matter from long-range transboundary air pollution.
42. WHO(2011). Air Quality and Health Fact sheet (313)
43. EEAThe European Environment, State and Outlook, (2010). Air Pollution.
44. EEAAir Quality in Europe, (2012). Report., 2012.
45. EEA(2012). Air Quality Standards (AQS):
46. EEAAir Quality in Europe, (2012). Report, 2012.
47. http://aeronetgsfc.nasa.gov
48. http://eceuropa.eu/environment/air/quality/standards.htm
49. http://wwwepa.gov/air/criteria.html
50. http://wwwwho.int/mediacentre/factsheets/fs313/en/index.html

Index

P

R

S

T

V

W